The

Canadian
Contemporary
Philosophy

Series

John King-Farlow
William R. Shea

Editors

Basic Issues in the Philosophy of Science

William R. Shea

Editor

Science History Publications
New York · 1976

First published in the United States by
Science History Publications
a division of
Neale Watson Academic Publications, Inc.
156 Fifth Avenue, New York 10010

First Edition 1976
Designed and manufactured in the U.S.A.

Library of Congress Cataloging in Publication Data
Main entry under title:

Basic issues in the philosophy of science.

   (Canadian contemporary philosophy series)
   Bibliography: p.
   1. Science--Philosophy--Addresses, essays,
lectures. I. Shea, William R.
Q175.3.B37   501   76-22198
ISBN 0-88202-160-5

Acknowledgement: The Editors gratefully acknowledge the secretarial assistance made available to them through the courtesy of Dean Walter Hitschfeld of McGill University.

Contents

Introduction: Contemporary Philosophy of Science   1
WILLIAM R. SHEA • *McGill*

Is Scientific Knowledge Rationally Justified?   15
THOMAS SETTLE • *Guelph*

The Hypothetico-Deductive Model of Scientific Theories: 36
A Sympathetic Disclaimer
ROBERT E. BUTTS • *Western Ontario*

Traditional Philosophy of Science: A Defense   58
JAMES W. VAN EVRA • *Waterloo*

The Philosophy of Biology   74
MICHAEL RUSE • *Guelph*

Neodarwinism, Mental Evolution,   91
and the Mind-Body Problem
THOMAS A. GOUDGE • *Toronto*

Scientific Objectivity and the Mind-Body Problem   107
JOHN T. O'MANIQUE • *Carleton*

Marxism, Social Science, and Objectivity   127
FRANK CUNNINGHAM • *Toronto*

The Relevance of Philosophy to Social Science   136
MARIO BUNGE • *McGill*

Towards the Assimilation of Rules   156
to Generalizations
ALEXANDER ROSENBERG • *Dalhousie*

Decision Making in Committees   172
ALEX MICHALOS • *Guelph*

Technology in Perspective   196
JOHN W. ABRAMS • *Toronto*

Notes on Contributors   210

To Ruedy and Emma

Introduction: Contemporary Philosophy of Science

WILLIAM R. SHEA

Philosophy of science in the second and third quarters of this century was dominated by a movement usually referred to as "logical positivism" or "logical empiricism." This school of thought took it for granted that science was a model for all rational discourse and that its theories were erected on the solid foundation of observational evidence. As we enter the last quarter of the century, winds of change are blowing at gale force. In the first essay in this volume, Tom Settle unleashes a lively attack on the assumption that science is a pre-eminently rational discipline and, in the second essay, Robert E. Butts offers a number of penetrating arguments to show that no observational language enjoys the unassailable status that the logical positivists claimed for it. The interpretation of scientific discourse offered by the logical positivists and known as the hypothetico-deductive model of scientific explanation is, nonetheless, still very much at the centre of the current debate on the nature of science. We shall be in a better position to enter this debate by briefly considering what the logical positivists attempted to achieve.

The advent of the theory of relativity and of quantum mechanics compelled philosophers of science to ponder how familiar language could be used to describe hidden entities of increasingly bizarre characteristics and what was really being referred to in this way. The logical positivists met this problem by retreating to the firm basis of the observable where there was thought to be comparatively no doubt about what people are talking about, nor about what our language means, nor about how we know our assertions to be true. As Rudolf Carnap put it in his "Intellectual Biography":

> We assumed that there was a certain rock bottom of knowledge, the knowledge of the immediately given, which was indubitable. Every other kind of knowledge was supposed to be firmly supported by this basis and therefore likewise decidable with certainty.

In other words, *the question of meaning* (What do theories mean?) and *the question of acceptability* (How do we know what theories are true?) were believed to hinge on *the question of reference* (What are theories about?) which, in turn, was taken to raise no difficulty. This methodological assumption—embodied in the classical distinction between self-evident "observation terms" and derivative "theoretical

1

terms"—constituted the first or *empirical* characteristic of logical empiricism.

The second feature, or the *logical* component, consisted in reliance on mathematical logic to formulate the problems of the philosophy of science, often rechristened "the logic of the sciences" to emphasize this aspect. Just as formal logic, since the time of Aristotle, had been concerned with the form rather than the content of propositions and statements, so philosophy of science was to deal with the logical form of scientific statements rather than their content. The job of the philosopher was conceived as the formal representation of scientific expressions in general while the task of confronting the results with actual scientific procedure was left to the practicing scientist. Important advantages were said to result from the adoption of the analogy of formal logic. Philosophy of science, thus disengaged from the specific content of particular scientific theories, became immune to change and the overthrow of current beliefs. Since the philosopher of science could, in principle at least, outline the characteristics of all possible explanations, he could also by the same stroke give the formal characteristics of all future explanations. This was one of the most attractive aspects of the logical empiricist programme: It appealed to the administrative machinery of our mind which is fond of classifications capable of development by deductive logic. Logical positivists concentrated on perfected or idealized systems and took for granted that science grew by the automatic incorporation of earlier hypotheses into later theories as special cases applicable in limited domains of experience.

The empirical and the logical aspects of logical empiricism can be summarized therefore in the following two tenets: Theoretical terms are (a) grounded in self-evident observation terms, and (b) open to exhaustive manipulation through modern mathematical logic. This unquestioning faith in the truth and in the obvious meaning of observational predicates was really a cover for a much more deep-rooted belief, stemming from Locke and the empiricist tradition: There are entities which are given in perception, and there are more or less well-defined areas of language which directly describe them. The positivist distinction between "observation" and "theory" is a direct lineal descendant of the distinction between "impressions" and "ideas" formulated during the development of British empiricism in the seventeenth and eighteenth centuries as a criterion of meaning and acceptability of concepts. But the "observational-theoretical" dichotomy differed from its predecessor in three respects. First, an attempt was made to avoid the misleading psychological overtones of the word "ideas" by speaking of "terms" and "statements." Secondly, Hume's

2

problem of the origin of such terms was avoided by justifying the acceptance or rejection of theories on the grounds of their "definability" and "reducibility." Thirdly, the new distinction was employed to tackle a problem to which the more general classification between ideas and impressions had not been applied. The modern distinction dealt with theories—a network of interrelated concepts—rather than with single ideas and had to consider rival sets of terms and statements. Hume and the early empiricists did not discuss competing sets of ideas, and it is here that the logical empiricists added a new dimension to the classical debate: They assumed that observational statements had direct empirical reference and were immediately decidable and saw, therefore, in the overlapping observational vocabularies of different theories a basis for instituting a comparison between them.

In view of the importance given to observational statements in assessing the relative merits of rival theories, it is surprisingly difficult to find any detailed explanation of their nature in recent works in philosophy of science. Accounts of the observational language are usually dependent on circular definitions of observability and its cognates, while the theoretical language is generally defined negatively as consisting of those scientific terms which are not observational. For instance, Richard Braithwaite writes in *Scientific Explanation*:

> ... experience, observation and cognate terms will be used in the widest sense to cover observed facts about material objects or events in them as well as directly known facts about the contents or objects of immediate experience.

Carl Hempel makes an analogous claim in *Aspects of Scientific Explanation*:

> ... an observation sentence might be construed as a sentence ... which asserts or denies that a specified object, or group of objects, of macroscopic size has a particular observable characteristic, i.e., a characteristic whose presence or absence can, under favorable circumstances, be ascertained by direct observation.

Even Ernest Nagel, who gives the most thorough account of the distinction in his influential *The Structure of Science*, seems to presuppose that there is nothing problematic about it:

> ... no precise criterion for distinguishing between experimental laws and theories is available, and none will be proposed here. It nevertheless does not follow that the distinction is spurious because it is vague, any more than it follows that there is no difference between the front and the back of a man's head just because there is no exact line separating the two.

3

The most explicit statement is found in Rudolf Carnap's classic paper "Testability and Meaning" where he makes the following two points: First, he holds that "observable" is a basic notion in terms of which other concepts such as "confirmation" and "testability" are defined because what it is to be counted as an "observable" is a question for psychology and the behaviourist theory of language, not for philosophy. Next, in approximate replacement of a definition, Carnap describes the concept of observability as follows: A predicate P is observable for a person N if, for some object b, N can under suitable circumstances come to a decision to accept or reject P(b) with the help of a few observations. He continues:

> This explanation is necessarily vague. There is no sharp line between observable and non-observable predicates because a person will be more or less able to decide a certain sentence quickly, i.e., he will be inclined after a certain period of observation to accept the sentence. For the sake of simplicity we will here draw a sharp distinction between observable and non-observable predicates . . . Nevertheless the general philosophical, i.e., methodological question about the nature of meaning and testability will, as we shall see, not be distorted by our over-simplification. Even particular questions as to whether or not a given sentence is confirmable, and whether or not it is testable by a certain person, are affected, as we shall see, at most to a very small degree by the choice of the boundary line for observable predicates.

Carnap presupposed, therefore, that an initial distinction between observables and non-observables on pragmatic grounds leads to no distortion in the discussion of the relation of theory to observation. But there is no such guarantee. This pragmatic account of observability overlooks that the pragmatic conditions themselves are always formulated against a background of current scientific views and opinions. In other words, they are themselves theory-laden. Furthermore, it is by no means obvious on Carnap's account that any predicate is, in principle, non-observable. Perception depends on training, past experience, and contemporary hypotheses, and there are circumstances in which suitably educated persons can come to a quick decision about any predicate, however apparently theoretical, such as "neutrino" or a "positron." The recourse to psychology or physiology to determine what any "observable" is offers no easy way out. Even if these sciences could provide us with a clear answer to the question, we would not escape the problem of what is to be accepted as an "observable" for psychology and physiology, and hence as an empirical basis for their theories of "observability."

The "firm basis" of observational language was becoming shaky, but the crisis was delayed by the possibility of shifting attention from the empirical to the logical side of the problem and concentrating on the deductive character of theories on the analogy of mathematical logic. Just as the terms "point" and "line" are not explicitly defined in a formalized geometry, but their meaning, or rules of use, is implicitly conveyed by the postulate set, so it was maintained that a physical predicate such as "positron" means just that entity which has the relation to other entities of atomic physics which are specified in the postulate system of physics. Physics differs from formalized geometry, however, in that some consequences of its formal postulates are translatable by means of correspondence rules (also called coordinating definitions, semantical rules, epistemic correlations) into observation statements, ensuring its empirical content, whereas a fully formalized geometry does not presuppose any empirical content.

Carl Hempel's "Studies in the Logic of Confirmation," which appeared in 1945, was the first major attempt to outline criteria of adequacy for the confirmation of statements related to evidence in deductive systems. Hempel enquired what general requirements ought to be satisfied by a confirmation theory which does justice to accepted forms of theoretical inference; he gave these two:

(1) *Special Consequence Condition.* If an observational report (referred to as E) confirms a hypothesis (referred to as H), then it also confirms every consequence of H. This seems desirable in order to allow for the familiar type of inference where a theory is advanced on the basis of experimental laws which it entails. For instance, Newton's theory is proposed on the evidence of Galileo's law of freely falling bodies and Kepler's laws of planetary motion which are entailed by it. Further as yet untested consequences are then drawn from the theory such as the variation in the period of a pendulum as it approaches the equator or the influence of the moon on the tides. These are predicted with assurance on the basis of the confirmation of the theory by its observed consequences. Indeed, confidence in the consequence of the theory is sometimes so great that it enables corrections to be made to laws that have already been accepted, as when Newton's theory amended Kepler's third law as it had been formulated on the observed evidence of the periodic time of the planets.

(2) *Converse Consequence Condition.* If an hypothesis H entails an observational report E, then the observational report E confirms any stronger hypothesis of which H is a consequence. This condition seems at first sight to be required since evidence entailed by a theory confirms the theory. But if it is taken together with (1), we get the following unsatisfactory case: Suppose H entails $E_1 \cdot E_2$, then E_1 confirms H by (2), hence E_1 confirms E_2 by (1). But suppose H is equivalent to $E_1 \cdot E_2$ (the mere joining together of E_1 and E_2), then

certainly H entails $E_1 \cdot E_2$; but there is no reason to expect E_1 to confirm any arbitrary E_2 which may be conjoined with E_1 to form H. Indeed, in general it will not do so. This example shows that if (1) and (2) are both accepted, any evidence will confirm any statement whatever by the mere construction of a suitable H by conjunction.

For this reason, Hempel originally rejected the Converse Consequence Condition and retained the Special Consequence Condition, thus demanding a stronger relation between hypothesis and evidence than mere entailment if evidence is to confirm the hypothesis. Carnap demonstrated, however, that this solution gave rise to serious difficulties, and Hempel has now accepted the Converse Consequence Condition and abandoned the Special Consequence Condition.

But whether a probabilistic theory is adopted or not, the upshot of the discussion is that the deductive criteria thus far suggested are unsatisfactory. It is hard to see how they could even begin to be relevant to the crucial problem of predictive inferences of the kind illustrated by Newton's theory and the laws it explains. For instance, let us suppose that we are given an axiomatic system (which we label S) from which some observation statements can be derived with an appropriate set of correspondence rules. Since S itself, on the deductive view, contains no observation predicates, no reason can be derived from the statements of S why one observation predicate should occur in a given correspondence rule rather than in another. On the one hand, if the conjunction of S and the correspondence rules is merely to explain given observations, then the observation predicates in these rules are determined by the known relations of observation predicates in observation statements. On the other hand, if S is to be used predictively, this involves deriving statements containing new observation predicates which do not appear in the correspondence rules. But S is powerless to determine what additional correspondence rules should be introduced, and consequently no predictions involving some particular new observation predicate rather than some other can be made with confidence from the theory. There are, however, countless examples of scientific inferences in which just such predictions—for instance, how the earth would appear when seen to astronauts landing on the moon—have been made with confidence.

In spite of its detailed analysis of the logical structures of postulate systems, the deductivist account of theories revealed itself grossly inadequate. A strong reaction was inevitable, and by the sixties there were increasing complaints that through its concentration on technical problems of logic, the logical empiricist movement had lost contact with real science.

It was naive to read the past, as the positivists did, as the record of great men throwing off the shackles of a dark inheritance to herald the

6

dawn of scientific objectivity. Many older theories that were allegedly laden with superstition, for instance, medieval mechanics and the phlogiston theory, were found to contain far more than simple-minded error and prejudice. It was discovered that Newton not only framed non-empirical hypotheses but was strongly influenced by the alchemical tradition, and that Galileo, the father of "empirical science," neither dropped balls from the Leaning Tower of Pisa, nor cared for experiments as much as had hitherto been believed. As closer attention was paid to the framework of theories, it became apparent that the theoretical context determined not only the questions that were raised but the terms in which the answers had to be expressed to be judged acceptable. This was a direct challenge to the logical empiricist position that there is an absolute, theory-independent observation language whose terms have the same core of common meaning for all competing theories. According to the new view, the meaning of all scientific terms, whether factual or theoretical is governed by the paradigm which underlies them and in which they are imbedded. This new interpretation has been urged with considerable vigour by Thomas Kuhn in *The Structure of Scientific Revolutions* and Paul Feyerabend in several articles and, more recently, in *Against Method.* "The meaning of every term we use," Feyerabend writes in "Problems of Empiricism," "depends upon the theoretical context in which it occurs. Words do not "mean" something in isolation; they obtain their meanings by being part of a theoretical system." Whereas the logical empiricists considered theoretical terms as wholly dependent on observational ones, from the new viewpoint the exact reverse is true. According to Feyerabend:

> The philosophies we have been discussing so far assumed that observation sentences are meaningful *per se*, that theories which have been separated from observations are not meaningful, and that such theories obtain their interpretation by being connected with some observation language that possesses a stable interpretation. According to the point of view I am advocating, the meaning of observation sentences is determined by the theories with which they are connected. Theories are meaningful independent of observations; observational statements are not meaningful unless they have been connected with theories. . . . It is therefore the observation sentence that is in need of interpretation and not the theory.

It follows that a basic shift in the theoretical viewpoint entails a change in what counts as a real problem, a correct method, an acceptable explanation, and even a fact, since the meaning of observational terms is determined by the theory in which they occur.

Kuhn and Feyerabend have mercilessly exposed the inadequacies of the usual formulation of the distinction between theory and observation, and they have shown that the concept of "explanation" in

7

science cannot be divorced completely from a consideration of the history of scientific explanations. In their reaction against logical positivism, however, they seem to have embraced an equally extreme view. For all its value and its suggestiveness, their position has not been formulated in a way that resolves the major problems of philosophy of science. It has only succeeded in making them more glaring—for if the meaning of every term depends on its theoretical framework, a change of theory must produce a change of meaning of every term in the theory. But this sweeping solution brings a new series of problems in its wake. Does the mere extension or application of a theory make a difference to the theoretical content and hence to the meaning of the terms involved? Does an alternative axiomatization alter the theoretical content so that the meanings of the expressions axiomatized change with reaxiomatization?

Apart from these technical difficulties, more fundamental objections can be raised: What, for instance, is the point of making experiments if they can be interpreted to support any theory? If the meaning of observational terms depends completely on the theoretical framework, how can we ascribe any continuity to the different usages of the same terms in successive theories? Rival hypotheses can no longer be said to contradict one another since in order for two sentences to be contradictory what is denied by one must be affirmed by the other, and this is meaningless unless they have something in common. If Kuhn and Feyerabend have established the bankruptcy of a philosophy of science based on the allegedly firm foundation of "observations," they have produced an alternative account that severs it from empirical evidence altogether. Their position eventuates in a complete relativism in which it becomes impossible to compare any two scientific theories and to choose between them on any but the most subjective grounds. In his incisive essay, James Van Evra shows that if observation itself shares the uncertainties of theories, we cannot avoid being cast adrift in a sea of hypotheses. Without a solid observational basis, we would seem to have no moorings, no plan, and no science.

Any viable account of the observation language must be able to show not only that we can keep afloat in such a sea, but that we can make progress through it. Several attempts have been made recently to plot such a course. One of the most promising is Mary Hesse's outline of a self-correcting confirmation theory in her book *The Structure of Scientific Inference*.

An essential feature of her model is that a predicate may be wrongly ascribed to an individual and that a definite small prior probability can be assigned to the occurrence of this error. Since this value is small, the ascription of predicates will for the most part be correct, and certain regularities in the relation between predicates will begin to emerge. For instance, it may be the case when a predicate has been applied to some individual, another predicate has usually been found to belong to it as well. Each predicate can be visualized as a knot in a network of relationships with other predicates where the strength of the strings between the knots stands for some increasing function of the proportion among individuals in which the predicates have been reported as co-occurrent or co-absent. The crucial problem is to detect and correct the erroneous ascription of a predicate. Since all evidence is necessarily in terms of reports that a particular predicate applies to a particular object, and it is assumed that there is finite probability that such a report may be wrong, a distinction must be drawn between a report and "what really is the case." In this situation one must try to identify errors on the basis of other reports, some of which may make the correctness of a given application extremely improbable. It is at this point that a charge of circularity could be brought against Dr. Hesse's model. She believes that it can be avoided, however, for while it is true that all predicates could be erroneously applied, not all could be known to be wrong at the same time: Alchemical terms used in modern chemistry, for example, are no longer predicated of individuals in the same sense as they were in the middle ages, but the transformation of their meaning did not happen for all of them at the same time. What is contrasted with the report that something or other is the case is not what really is the case, to which we have no direct access, but rather what in a given theory should be the case, which may in a small proportion of cases contradict what has been reported. That this is not as arbitrary as it may seem can be illustrated by the following case-history in geology: When Louis Agassiz, fresh from studying glacial action in Switzerland, visited Scotland in 1840, he immediately recognized moraine, roches moutonnées, and glacial striae as distinctive features of the Scottish scenery. In this case, his extensive knowledge of the effects of glaciers and glaciation may be said to have enhanced his perception. But this way of looking at objects became so ingrained, that he later began to *see* evidence of glacial action where certainly there had been none, as in the Tijuca hills behind Rio de Janeiro. His reports, however, were only revised when it was shown that they were *inconsistent* with later and more detailed accounts.

It does not follow from this that the theory which contradicts the fewest reports is the theory with the highest confirmation value: The amount of "correct evidence" according to this theory may be counter-acted by the higher prior distribution of some other theories. For instance, preference can be given to theories which cluster individuals, that is to say, which allow individuals to be classified in more or less well-defined classes with high similarity between members of a class and a non-member of that class. Without introducing further possible refinements, we can see how this model, however oversimplified, offers an alternative to the positivist observational language.

On this new interpretation, the meaning of descriptive predicates is no longer determined exclusively by direct empirical reference in a way that is radically different from theoretical predicates. The meaning of both observational and theoretical terms has elements of stimulus-response in situations of easy empirical reference and also of contextual relation with other predicates which co-occur or are co-absent. It follows that whereas error on the positivist view is taken to be relative to "what is really the case," on this interpretation it is considered relative to a theory. This has the interesting consequence of allowing the possibility that the source of error may not merely be lack of attention to what is observed, but can also occur in a sincere and careful report, when ascription of a predicate does not fit in well with the general outline of the theory. This is to say that the correct ascription of a predicate depends both on a careful response to a stimulus situation and on the relations of this predicate to other predicates in this and other individuals. A set of such relations constitutes a theory. Hence confirmation and falsification will depend not on one allegedly crucial experiment, but on all the reported evidence in other situations, together with the verdicts of different theories on the correctness of these reports. Thus, the directness of empirical reference and the immediate decidability which the positivists considered as those characteristics of observation predicates which mark them off from theoretical predicates appear illusory.

Whether this approach suggested by Mary Hesse can be shown to be adequate to all types of theoretical inference is too extensive a question to prejudge here, since it requires a confirmation theory based on a language containing the real number continuum, for which many as yet unsolved mathematical problems arise. But perhaps enough has been said to conclude that we are not compelled to adopt a kind of historical relativism and to regard theoretical inference as fundamentally irrational. The problems and the techniques of the logical empiricists can be pushed further than they themselves thought possible. Both the question of meaning and the question of confirmation are more complex and interrelated than had hitherto been imagined; but there is, as yet, no reason to despair of a logic of scientific discoveries.

Kuhn and Feyerabend are right in stressing that the framework of contemporary hypotheses, the paradigm or "disciplinary matrix," determines to a large extent what questions can be raised and what views can be suggested about a given scientific problem. They fail to explain, however, how different theories can be contrasted and appraised. For instance, compare the medieval theory of impetus with Aristotelian mechanics on the one hand, and with the principles of inertia on the other. The impetus theory shares with Aristotelian physics the view that a constant force is needed to keep a body in motion, but it tends toward the modern view by making this force internal and incorporeal rather than external and corporeal. It can be argued that the impetus theory encouraged a fresh approach to traditional problems by removing long-standing conceptual barriers. The Aristotelians rejected outright any suggestion that the earth rotated because the physical assumptions of their system entailed that a strong wind would be set up in the direction opposite to the earth's motion. But the impetus theory made it possible to entertain the idea that the earth might move by affirming that the air would receive an impetus and would be carried around as though nothing has happened. Likewise, by internalizing the cause of motion it shifted attention to a new set of possibilities. Since air was no longer the force producing motion but merely an impediment, motion in a vacuum ceased to be impossible, and the ground was cleared for thinking about the idealized case of a body moving in the absence of impeding forces. Furthermore, by treating all cases of motion, whether terrestrial or celestial, natural or constrained, in terms of one kind of cause, namely impetus, it paved the way for a unified account of all motion and provided an alternative approach to the traditional Aristotelian division.

On the Kuhn-Feyerabend view one is practically driven to describe scientific change in revolutionary terms, to speak, for instance, of the overthrow of Aristotelian mechanics by the impetus theory and of the latter by Newtonian science. A consequence of this position is that theories are "incommensurable" and that change is ultimately irrational. A more balanced description can, however, be attempted. The Aristotelian-Scholastic tradition applied, as a template, an intricately connected web of concepts and propositions to the data of perception and everyday experience. In laying down this system of interpretation, it simultaneously set up obstacles or limitations, both by theoretical precept and by suggestion, to thinking in certain other ways. The vacuum and the actual infinite, for example, appeared self-contradictory, while the motion of the earth seemed physically impossible. Scientific advance is related to a reassembling of the pattern of our experiences, and there is immense resistance to this. In this sense the Aristotelian view involved certain presuppositions that specified what could and what could not count as an explanation. But these presuppositions could be recognized as such and, in fact, they were identified and rejected by Philoponos, Oresme, Buridan, and others, who reassembled the facts in a different pattern!

The relativism of the Kuhn and Feyerabend position is not the result of an investigation of actual science and its history, but merely a logical consequence of a narrow presupposition about the "meaning" of scientific terms. They hold that if the terms do not retain precisely the same meaning over the history of their incorporation into more general theories, then these theories cannot be compared, and the similarities they exhibit must be considered at the best as superficial and at the worst as deceptive and misleading. This claim rests on the assumption that two expressions or sets of expression must either have exactly the same meaning or must be completely different. The only possibility left open by this rigid dichotomy of "meanings" is that history of science, since it is not a process of development by accumulation, must be a completely non-cumulative process of replacement.

The inherent weakness of this position turns out to be its retention of a positivistic concept of "meaning." If anything, the revolution isn't radical enough. Kuhn and Feyerabend, in spite of their spirited attack on the positivist view that "theories" are parasitic on "observations," nevertheless approach their problems with that distinction in mind. They have applied the old classification to a new purpose rather than invented new ways of dealing with old problems. They have merely inverted the respective roles of the two members of the classical distinction: It is now the "theory" that determines the meaning and acceptability of the "observation," rather than the other way round.

It is much more radical to question the distinction itself. It may well be that some problems are heightened (if not created) by the deficiencies of the distinction between a theoretical and an observational language. If this is the case, it is no longer the solution that is seen to be problematic, but the way in which the question is framed. For instance, the notorious problem of the status of theoretical entities or the question whether a realistic interpretation of scientific theories can be upheld may be partly generated by an inadequate concept of the working of theories. It is all too easy to view the distinction between observational and theoretical as paralleling a distinction between existent and non-existent. If observation terms are said to have a clear and direct reference to entities that exist while theoretical terms do not, it becomes difficult to know what theories are about. This is not due to any intrinsic opacity of the concept of existence, but to the sharpness of the distinction between theory and observation. As long as it was believed that theoretical terms could be exhaustively described by observational ones, theories could be handled as a convenient short-hand. When it became apparent that such a reduction was only possible in part, the extra meaning of theoretical terms was sought in the position they occupied in the context of the system to which they belonged. Thus, theoretical entities could not exist in the same sense as tables and chairs, and it became a first-class puzzle to know in what sense they could now be said to exist, short of being merely useful fictions.

The reaction of writers such as Kuhn and Feyerabend was to claim that this indicates that observations are governed by theories which are, in the last resort, irrational guesses at what the universe really looks like. But this did not solve the problem of meaning: It merely replaced the positivist thesis of meaning invariance with the doctrine of incommensurable meanings. An alternative is to consider "meanings" as similar or analogous: comparable in some respects while differing in others. By taking this path, we can hope to preserve the fact that theories, such as Newtonian and relativistic dynamics, are not incommensurable although they are profoundly different.

The difficulty in this interpretation lies in the concept of similarity or degrees of likeness of meanings. It is only too tempting to distinguish, against the background of a particular theory, between what is and what is not an essential part of the meaning of a term. It seems obvious, for example, (in the light of subsequent developments) that Newton's absolute space and absolute time are "irrelevant features" of his mechanical theory. Yet those very features, for some purposes, may prove to be the very ones that are of central importance in comparing two uses. An absolute distinction between essential and accessory features could only rest on a conception of scientific understanding that presupposed the answer to what is being queried.

The alternative to the hypothetico-deductive model will have to be both faithful to the actual history of science and adequate to the requirements of the logic of theoretical inference. The discussion, which is only in its early stages, will involve a critical examination of the assumption that scientific inferences rest on analogy, and, if so, in what sense. Whatever the final outcome of the debate, the benefits already accrued from the radical reappraisal of the observational-theoretical distinction make the venture one of the most promising in contemporary philosophy of science.

While the quest for a successor to logical positivism continues, the hypothetico-deductive system remains a model that can be used profitably in clarifying by contrast the nature of the problems that confront philosophers of more recent disciplines such as biology and social science. Michael Ruse, in his lucid essay on the philosophy of

13

biology, shows how the theory of evolution can be described in the language familiar to philosophers of physics and why this approach is not entirely satisfactory. Alexander Rosenberg performs a similar operation when he compares rules established by social scientists with generalizations made by scientists in other fields.

John O'Manique discusses the mind-body problems in the light of the history of philosophy and physics, and Thomas Goudge argues persuasively for the possibility of making headway in the investigation of this problem by using the conceptual tools provided by neo-Darwinsim. John Abrams addresses himself to the question of technology and sets it in a broader philosophical perspective than is usually done.

Frank Cunningham enters an important reminder of the importance and sophistication of Marxism in dealing with the problem of objectivity in the social sciences. Mario Bunge, in a provocative essay, argues that the social sciences stand in need of an exact and rigorous philosophy, and Alex Michalos illustrates how the tools of the logician are relevant to decision making in a world of proliferating committees.

Bibliography of Works Cited

Braithwaite, Richard C., *Scientific Explanation.* Cambridge: Cambridge University Press, 1959.

Carnap, Rudolf, "Testability and Meaning" in H. Feigl and M. Brodbeck (eds.), *Readings in the Philosophy of Science.* New York: Appleton-Century-Crofts, 1953.

————, *Logical Foundations of Probability.* Chicago: Chicago University Press, 1962.

————, "Intellectual Autobiography" in P.A. Schilp (ed.), *The Philosophy of Rudolf Carnap.* Lasalle, Illinois: Open Court, 1963.

Feyerabend, Paul, "Problems of Empiricism (Part I)" in Robert G. Colodny (ed.), *Beyond the Edge of Certainty.* Englewood Cliffs, N.J.: Prentice Hall, 1965.

————, *Against Method.* London: New Left, 1975.

Hempel, Carl G., *Aspects of Scientific Explanation.* New York: The Free Press, 1965.

Hesse, Mary, *The Structure of Scientific Inference.* London: Macmillan, 1974.

Kuhn, Thomas, *The Structure of Scientific Revolution.* Chicago: Chicago University Press, 1962.

Nagel, Ernest, *The Structure of Science.* London: Routledge and Kegan Paul, 1961.

Is Scientific Knowledge Rationally Justified?

TOM SETTLE

1. Introduction: How Basic Is This Issue?

The question with which this essay is entitled may seem easily answered in the affirmative: One might say that if what a scientist thought was knowledge turned out not to be justified, it would not be knowledge. Or one might say that surely some knowledge is rationally justified, and if any knowledge is so justified, knowledge of scientific laws and of the results of objective tests is. If either of these responses is on the mark, the question in the title is dealt with and that is that. But I shall try to show that there is more to the question than meets the eye: that whether scientific knowledge is rationally justified is a deep problem. And I shall argue that the usual and obvious solution is in error. If I am right, there are important repercussions that scientific knowledge lacks a final *rational* justification, and that in so far as it is justified, it is a matter of social consent rather than proof. Science's air of being rationally justified is explained by there being a rational component to social standards of justification, standards which always fall short of guaranteeing the truth. This rational component is commonly misunderstood.

I think it can be said fairly that the issue raised in this essay is basic to philosophy of science, which has grappled with the problem of justifying scientific knowledge throughout the modern period beginning with Bacon and Descartes, not to mention earlier debates. Moreover, this issue is basic to social and political affairs, since if any justification of knowledge in science may ultimately be a matter of social consent, scientific knowledge, including knowledge in economics and political science, can hardly be used to replace social consent in deciding social issues. There is another important sense in which this issue is basic to all rational inquiry in the sciences as well as in philosophy: Rational justification has long been thought of as a method for distinguishing what is true from what is false, and so long as it was thought that we could justify some human knowledge rationally, the pursuit of truth could be combined with or even replaced by the pursuit of justification. If we can never have a rational justification which guarantees truth, however, scientists might seem to be faced with several important choices: whether to abandon the search for truth and search only for what can be partially justified rationally or to look for an alternative method of pursuing truth, with or without guarantees, and with or without rationality. But this is not

15

quite the case. Let us assume that an aim of doing science is to find truth and that the method used is rational (as it has always been thought to be). Perhaps what is needed is not so much a radical re-shaping of science as a replacement of the popular view of it. Perhaps many scientists have been doing all along something rather different from the account philosophers have given of their activity. It is possible to construe scientific knowledge as conjecture rather than as justified true belief, and to construe testing in science as aimed at criticizing theories rather than confirming them as has been suggested by Popper.

The main thrust of the paper will be to show why a negative answer has to be given to the title question and to discuss some implications of disagreeing with the popular view that scientific knowledge is rationally justified. Accepting that the popular view is wrong may leave room for a nonjustificationist theory of rationality, or it may lead thinkers to supplement reason with some other justification, such as the authority of experts. I shall discuss both these alternatives and why I prefer what I call a skeptical theory of rationality.

If "truth" be understood broadly to include not only what is so about physical reality but also what is so about ethical obligation, it is easy to see how fundamental the issue of the justification of knowledge is in practical affairs. That we may not attain unanimity in its resolution does not diminish the importance of the problem. Unfortunately, there will not be space here to take up all the questions whose link with the subject of this essay gives to it its fundamental importance. I can do little more than hint at some of the issues involved, in passing.

2. The Failure to Give a Rational Justification of Scientific Knowledge

Let us assume for the moment that scientific knowledge is a set of sentences intended to describe reality. (For other purposes this definition may need amplifying to include coherent intuitions of reality, as argued by Polanyi and Hanson, but I excuse myself from introducing what would only complicate discussion at this juncture by pointing out that the amplification would only make stronger the case that I wish to adduce against the idea that scientific knowledge lacks rational justification, and by promising to return to the matter in Section 6.) And let us further assume that some of these sentences report particular matters of fact, such as that an event with some particular characteristics took place at a specific time and place, while others assert

16

generalities about reality, such as that any event with certain characteristics would be followed immediately by events with certain other specific characteristics or that any object of a certain kind would behave in a specific way if placed in circumstances of a certain kind. Examples of particular sentences in science are: the length of time a particular bean took to sprout after being placed in certain surroundings; the amount by which a particular rod expanded in length on being heated through a specific temperature difference; the score a particular child made at a particular age in a specific standardized test. Of course specific scientific knowledge is rarely stored in fully expressed sentences. More often than not tables are used as shorthand collections of specific reports of observations. Some complex matters of fact may be stored photographically, for instance the result of taking a cell-slice may be recorded on film using an electron microscope, or a shot may be taken of the tracks left in a cloud chamber by a cosmic ray.

Despite that the bulk of scientific knowledge is undoubtedly a collection of reports of experience, variously expressed in descriptive sentences or tables or photographs or other records, an indispensable component of scientific knowledge is undoubtedly the set of general or universal sentences which comprise scientific theory. These are of two importantly different kinds: generalities which are very close to experience and may look like a summary of a series of reports; and generalities more remote from experience which introduce entities (such as electrons) or properties (such as change of energy-state of electrons) which are never a matter of immediate observation but which are invoked to explain matters of direct observation. The change of energy-state of electrons, for example, may explain absorption and emission of light. Let us call the first kind of generalizations empirical laws. Some branches of science seem rarely to go beyond empirical laws in their descriptions of reality, while other branches, such as physics, seem to be comprised nowadays primarily of theories, most of whose components are remote from observable experience.

Perhaps readers should be warned that the term "theory" is rather ambigious both in common scientific usage and in the usage of philosophers of science, let alone in ordinary use (consult the *Shorter Oxford Dictionary* or *Webster's*). There is a two-fold ambiguity: First, "theory" sometimes refers to a whole body of knowledge in science and sometimes only to its key sentences. For example, if the theory of gravitation is expressed fully, it will include presuppositions, axioms, postulates and interpretative sentences as well as the key relation

$$F = G \frac{M_1 \, M_2}{r^2}.$$

17

For brevity gravitation theory is sometimes said to be just that key relation. Secondly, knowledge thought secure is sometimes called "fact" and contrasted with knowledge not thought secure, called "theory." This is a very confusing contrast, not only because what is thought theoretical because general will differ from what is thought theoretical because speculative, but also because of the confusion over the use of "fact" and because of erroneous ideas about the security of knowledge.

To view science as a set of sentences intended to describe reality either particularly or in general allows us to draw an important distinction which removes some of the ambiguity from use of the word "fact" common among scientists. I shall retain the strict sense of "fact" found in the *Shorter Oxford Dictionary* (sense #3): "something that has really occurred or is the case," and thus distinguish facts (occurrences) from reports of facts or descriptions of facts, which comprise the particular or singular sentences of scientific knowledge. Then we can say that a sentence is true if what it reports is so—if, that is, it corresponds to the facts. Theories will be true if the real world is orderly in the manner they describe.

These clarifications dealt with, it is now possible to indicate what kinds of rational justification have been attempted for scientific knowledge. There are two main kinds of attempts which I shall call respectively intellectualism and empiricism. In the context of this discussion, we may say that intellectualists attempt to justify knowledge from the intellect and empiricists attempt to justify it from experience. In the modern period, the earliest intellectualist account of science was Descartes' and the earliest empiricist account was Bacon's. Both accounts have been modified and refined as a result of criticism, and have appeared in a variety of versions, of which I shall indicate only the most famous. The varieties are most easily distinguished by pointing to the problems each attempt encountered and the different strategies that were adopted to solve or to side-step the problems.

Descartes' account of science was a deliberate attempt to give it the kind of certainty he thought mathematics had by having scientists adopt the methods he thought mathematicians used. Descartes made significant contributions to mathematics (for instance, in algebraic geometry) and was familiar with the method of deductive proof of theorems, given axioms and postulates. His programme for science required the discovery of the fundamental axioms and postulates that governed the real world, followed by deductive derivation of less universal generalizations, and even—given the assumption of spatial and temporal coordinates—specific statements of fact. It was an exciting and bold programme which foundered at least in part because

18

of the difficulty of discovering any undeniable axioms as starting point. Science was not without candidates for undeniable axioms ("clear and distinct ideas" in Descartes' terminology) but none gained universal approval for long. Even Euclidean geometry, which held more or less undisputed sway for two thousand years, fell when it was discovered that geometries could be constructed in which Euclid's postulate about parallels was denied. (See the work of Gauss, Lobachevsky, and Riemann, for example.) Although Descartes' position has been more or less completely abandoned as far as the justification of scientific knowledge is concerned, his analysis of the logical connection between theories and reports of experience has been retained.

Kant made an interesting attempt to give intellectualist certainty to scientific knowledge by claiming that the categories which governed inquiry (space, time, causality, and so on) were categories of thought itself, and thus that it was psychologically impossible for scientific inquiry to proceed on any other basis. The price he paid for the certainty he thought he could give science was very high, however. Kant drew the distinction between objects in the real world, things-in-themselves or noumena, on the one hand, and objects as they appeared to us in observation, or phenomena, on the other hand. He then claimed that there could be no human knowledge of the noumena. Thus, scientific knowledge was restricted to knowledge of the phenomena, knowledge of appearances, knowledge of what we might call the observable world. There is still a debate today between those who agree with Kant that scientific knowledge is only of phenomena—some hold that there is no noumenal world; others, that if there is, its properties are of no importance to us—and those who hold that scientific knowledge is an attempted description of the noumena, of the invisible world. While the latter have had to abandon the strict requirement that theories in science be rationally justified, it is hard to distinguish the former's theories of scientific justification from those of certain empiricists.

Bacon assumed that while the aim of science was to understand and control nature, the starting point of inquiry was observation. The crux of Bacon's account of science was his idea that after a lengthy and painstaking collection of empirical data, it would be possible to arrange the data in tables which would suggest empirical laws, and that groups of empirical laws would suggest more abstract principles. This method of inferring generalities from particulars is called induction. It was suggested by Bacon both as a fruitful method of inquiry and as a satisfactory method for justifying the resulting general principles. Let us leave aside whether or not induction is a fruitful heuristic, and ask merely whether a general law suggested by a collection of data with which it is consistent is justified by these data.

19

Long before Bacon's day, doubt had already been cast on the possibility of secure knowledge of generalities given any number of instances both by Aristotle, who argued that the truth of a generality was not guaranteed by the truth of instances of it, and by Maimonides, who argued that there are not any grounds for supposing that the Laws of Nature stay unaltered through time. But the death-knell of inductivism in the modern period was sounded by Hume who, confusing the two objections from earlier thinkers, argued that past consistent behaviour did not guarantee any future repetition. Kant himself understood his attempt to secure scientific knowledge as an attempt to meet Hume's objection. Kant claimed that the order phenomena seem to possess is imposed upon the phenomena by the intellect. Empiricists did not think that Kant dealt adequately either with Hume's objection or with other problems which faced a rapidly growing science—what status, for instance, to accord to theoretical entities, such as invisible particles or fields of force, and what status to accord to theories couched in those terms. Was Faraday's concept of electromagnetic field a mere fiction, useful for inquiry? Was Maxwell's theory of electromagnetism just shorthand for a collection of data? Well versed as Kant was in the physics of his day, he could hardly be expected to have anticipated the problems posed by later revolutions in physical theory. The combined effect of Hume's criticisms of induction and Kant's claim that the noumenal world was beyond sensual perception was to make way for the view that scientific knowledge consisted entirely of particular statements of fact about phenomena. Scientific theories were thought mere instruments for prediction or for the storage of unwieldy bodies of data.

Consistent with this view was the verification principle adopted by a group of philosophers and scientists in Vienna (the Vienna Circle) whereby only those sentences which could be verified in experience had meaning. "Positivism" is a general name for the versions of empiricism which give such primacy to reports of experience. For positivists, the major problems in giving an account of scientific knowledge were in accounting for empirical laws and for theories. It was thus a major task for positivism to construct an inductive logic that could answer Hume's objections by yielding a measure of the degree of inductive support a generalization had from a body of evidence with which it was consistent. A second problem was that of the meaningfulness of theoretical statements which employ terms remote from experience. If theoretical statements lacked meaning because they lack verification, it was difficult to understand how they could at the same time be regarded as explanatory. Thus a series of urgent problems clustered together for the empiricist: the logical structure of scientific knowledge;

the meaningfulness of statements of different logical kinds; the support evidence gives to theories; what explanation is. As far as I can tell from the literature, none of these problems is yet resolved to the satisfaction of positivists, despite the best efforts of such leading thinkers as Carnap, Hempel, and Nagel. Let me mention a few other problems which are often assumed to be satisfactorily solved, but which in my view are insoluble. That they are insoluble is devastating for the point of view in philosophy of science that scientific knowledge enjoys rational justification.

3. More Unsolved Problems for a Justificationist Theory of Scientific Knowledge

In the presentation of these problems, I shall take it for granted that we can distinguish conceptually between the following: (i) the real world; (ii) the commonsense world, or public space; (iii) each person's sensual world (visual field filled out by the other senses), or private space; (iv) each person's perceptions. It is important philosophically to make these distinctions, though the importance may seem minimal in everyday life. Even so, it is commonplace for people to distinguish between their private sensual world and public space. When, for example, I put on my reading glasses or use a telescope, or when I strain to hear a faint noise, or sniff at a suspicious scent, I am aiming to alter both my perceptions and my sensual world, without any intention of altering the dimensions or the position or the properties of objects in public space.

What I am here calling the commonsense world, or public space, is distinguished from anyone's private world by being detached from particular viewpoints, but it resembles it in that those properties we think that objects in public space possess are just the properties we can perceive or think we could perceive if we were suitably placed (and wearing the right spectacles!). On the other hand, there is an irreducible theoretical component both in our perceptual world and in the commonsense world. In our private world what we expect things to be like and even what we wish they were like influences how they appear to us to be. With respect to public space, our knowledge of what it contains is derived not only from what we have perceived but also from a belief that some things remain unchanged or change at known rates, aided by a memory of how they once were. Furthermore, our knowledge of public space gets "corrected" in conversations with fellow humans who discuss with us how to interpret different visual, tactile, etc. signals. For example, if we are accustomed to the appearance of wet roads, we may mistake the refraction of light from the sky in a patch of hot air over hot roads for reflection in a wet patch. More experienced travellers will put us right. If we learned to fish when young, we may have noticed the difference between the real and the apparent depth of fish in a clear stream. Nonetheless, we could make the adjustment to our knowledge of public space even without the experience.

For the purposes of philosophy of science, the distinction between the real world and the commonsense world is of considerable importance because the commonsense world is what is commonly believed to be the real world. The commonsense world is what Kant called the phenomenal world: In his theory, it is a mental construct, whose verbal description is conventionally agreed upon. In my terms, it is a generalized or objectivized version of each person's private world.

The distinction between the commonsense world and each person's private world is further drawn by noting that it may be allowed that objects in public space possess properties some people, but not others, can perceive. For example, people with impaired vision or impaired hearing know that others can detect qualities in sight or hearing which they cannot.

For ordinary purposes of everyday life, it is usually taken for granted that human beings directly and correctly perceive reality. Of course, this naive realism leaves illusions unexplained, and the explanations to which we resort challenge the ordinary assumption that all of us correctly perceive the real world. No doubt our perception corresponds sufficiently closely to reality for us to be able to move about without injury most of the time, but this closeness of correspondence is not sufficient for the purposes of scientific inquiry, which is aimed at least partly to solve puzzles thrown up by ordinary perception—the illusion of the wet road on a hot day, for example—as well as to answer questions relating to the causes of everyday occurances.

In this essay, I shall take realism for granted: I shall assume that there is a real world independent of our experience and that our experiences are in some way caused by it. From this point of view, the commonsense world is a mental construct whose composition is given partly by theory, partly in perception. As the example of illusion suggests, theory sometimes corrects perception or, more strictly, ill-interpreted perceptions. The scientific enterprise can then be seen as an attempt to improve the theories by which our perceptions might be interpreted or explained. There is a very old philosophical debate whether the seeming greenness of a tree is a property of the tree or only of the perception of the tree. "Is a tree green when no-one is looking?"

one might ask. The answer given from the point of view I am here taking for granted is that while greenness itself is not a property of trees the propensity to seem green to human observers is (this propensity being a complex of the properties of leaves and of human eyes etc., with respect to the emission and absorption of electromagnetic radiation of a certain wavelength). Nonetheless, the problems I shall point to exist regardless of whether scientific theory is construed as referring to a real (noumenal) world or merely to a public phenomenal world.

Leaving aside the problem of induction and related problems concerning theoretical science which I have already mentioned, let me now draw attention to several insuperable problems whose failure to be solved undermines the justificationist position.

First there are two problems in correctly describing our experiences: The first is caused by our juxtaposing a conventional language to a private experience. Since no one but ourselves experiences our perceptions, it is a matter of guesswork whether the words we use convey correctly what we have perceived. And we have no way of checking the concept occurring in other people's minds as a result of our report. I am not saying that this is a serious practical problem in ordinary life—for the most part people seem not to encounter difficulties— but it is a serious problem if a justification of scientific knowledge is at stake. There cannot be a rational guarantee that I have correctly reported my private experience in conventional language.

The second problem stems from the characteristic that verbs and general nouns have of being general. No scientific law could be formed if reports of experience employed only specific nouns (proper names) or if each action was to be described with a specific verb. Indeed, significant conversation would halt. However we come by it, each of us seems to have a mental trick of classifying objects and incidents which is mirrored in ordinary language: Verbs name types of action and nouns name classes of objects. Thus the reports of particular experiences already take for granted a theory of the world as divided into kinds of objects and of properties. The classifications thus presupposed may get revised in the light of experience, of course. But no experience could be described in a conventional language without presupposing a theoretically ordered world (real or apparent). This poses a problem for a justificationist since he has to presuppose a fundamental taxonomic theory before he can use the particular sentences he wishes to use to describe his evidence for scientific theories. This problem compounds the problem of induction.

Further unsolved problems appear in a person's move to use either his private world or his understanding of public space to correct his perceptions. Both the public space and the private space of an individual include many items which are not directly perceivable such as the legs of the man behind the desk or the silent engine in the car with its hood down. These items may be supplied by memory, expectation, or some such. Neither memory nor expectation can guarantee the existence of the hidden items, however, even if perception were able to guarantee the existence of the perceived items. Hence neither the composite private sensual world our mind constructs nor the commonsense world of convention can serve as an adequate guarantee of reports of experience any more than perceptions can, despite its being coherent and organized.

Furthermore, if the chain of commonsense justification of scientific knowledge is to be complete, either the world of public space must be identified with the private sensual world, or the two must be infallibly connected so that a person's reports of what he sees may justifiably be regarded as reports of what there is to see. I know of no way to make this connection rationally justified, and I think that the existence of illusions is a sufficient refutation of the possibility of making it, as Robinson has argued. The point may be put this way: Whenever we suffer an illusion which we later correct, we have had to correct an item in our sensual world, while public space has remained unchanged. Hence either before or after the correction our sensual world did not correspond to the apparent (phenomenal) world.

As I have said, these serious problems exist for justificationists regardless of whether scientific theory is construed as referring to noumenal or a phenomenal world. And there is no escape by saying that science is merely a collection of reports of specific occurrences because these reports will be couched in terms which presuppose an unjustified taxonomy of objects and events in the world under investigation.

Since I have allowed that many of these problems lack significance for practical purposes most of the time, it might be wondered why I am being so strict in the demands I make on the justificationist position. Let me consider next why *partial* justification is not enough.

4. Why Partial Rational Justification Is Not Sufficient

Whether partial justification is good enough depends on our purpose. If we have purely practical purposes in mind, it may suffice for some particular theory that known arguments in favour seem to be stronger then the arguments against. Or we may decide we cannot afford to spend more time and money trying to improve a theory; or we may like taking risks. But these practical purposes are beside the point if we are asking whether scientific knowledge is true. We may, of course, define

24

the purpose of doing science as the development of that set of theories modelling reality which we can best justify according to some standard of partial rational justification. In that case, partial justification is obviously good enough. The main problem might then seem to be finding a satisfactory measure of partial rational justification. But this would be misleading: An even more serious problem would be to justify any proposed measure of partial justification as a satisfactory measure in some sense of satisfactory that went beyond the merely subjective. If there are several candidates for measures of partial rational justification, is there a method by which we could justify our choice of one rather than the other which did not lead to an infinite regress, did not involve circularity, or did not change the basis of justification from rationality to something else? I suspect not.

Probability was for long a live candidate as a measure of partial justification. But it proved very difficult to develop a theory to measure the probability of an hypothesis on certain evidence that would rank hypotheses as to preferability in an order that corresponded to the intuitions of working scientists. Even if we could have satisfied their intuitions that probability really did select the best theories, that would have been merely a partial nonrational justification of the criterion of partial rational justification. Of course, if that particular candidate had secured a high proportion of the votes of working scientists, that would have turned the votes of working scientists into the criterion of partial justification instead of probability.

Perhaps we could have used the votes of working scientists more directly to decide matters of truth. Polanyi proposed that the decision of a qualified scientist was the nearest we could get to truth. He based this on a theory of connoisseurship as the best warrant for a truth-claim. Ziman and Kuhn both took up the idea and developed theories of knowledge or of growth of knowledge in which consensus within the scientific community plays a crucial role. Such moves, however, constitute abandoning rational justification as a criterion of truth. The ideas of these thinkers are very attractive, but we should notice that they all explictly renounce the use of a strict rational justification. In this respect I agree with them; but I do not accept their alternative proposal, which is to be satisfied with a non-rational justification, with reason taken as far as it will go. I shall give more attention to that alternative in section 6.

If, however, instead of supposing that science seeks merely those theories that can be partially justified, we suppose that the purpose of doing science is the development of a set of *true* theories, then the pursuit of justified theories seems justified only so long as justification guarantees truth. But partial rational justification does not even purport to guarantee truth. Hence the pursuit of merely partial justification is not the immediately obvious route by which to pursue truth. The question how best to pursue truth needs to be opened afresh: The failure of both the intellectualist and the empiricist programmes to deliver a methodology which guaranteed true laws or theories raises once again the questions of how knowledge grows in science and what its status is.

Before discussing two major rival candidates, let me give one further reason why partial rational justification is not good enough. Much of the technology upon which people daily or occasionally rely, and whose failure causes deaths, sometimes many deaths, presupposes scientific knowledge. Of course, each technological innovation before being implemented needs to be challenged both as to what its effects will be if it works and if it fails. The prediction of the effects of implementing this or that novelty in society relies completely upon scientific knowledge: Hence the judgment whether to implement, in so far as it hinges on a comparison of possible effects of implementation with possible effects of non-implementation, relies on science. It becomes of immense social consequence whether scientific knowledge is reliably right. Sometimes, of course, scientific knowledge is not right—and many lives have been lost in consequence of a misplaced trust. Let us suppose that no trust in scientific knowledge would ever be misplaced if scientific knowledge were fully rationally justifiable: It needs to be asked what social standards of justification are called for if rational justification is always only partial. Surely a government would not have fully discharged its responsibilities to its citizens if it let technologists implement any novelty for which there was only a partially rationally justified assurance of non-deleterious consequences. In addition, a government would seem obliged to consider the price of the assurance's being wrong. Thus partial rational justification of scientific knowledge is not good enough for practical purposes, especially when the cost of error is high.

5. Rationality without Justification

I begin this discussion by drawing attention to what I think is a serious mistake in the majority of accounts of the rationality of science and suggesting how to avoid it. Then I should like to point out some consequences of adopting a different view of the rationality of science. Undeniably, a reasonable person in choosing what to believe (in so far as anyone chooses his beliefs) will pay attention to the comparative weight of argument for and against rival ideas. Similarly, in choosing how to act, he will weigh the consequences favourable and unfavourable to himself and others of various alternatives. (Let us assume he wants to believe what is true and do what is right.) The mistake to which I refer is identifying weighing arguments for and against a belief (or an action) with justifying it. Without digressing into a history of the usage of the term 'justify,' I want to point to various ambiguities arising from the differing standards, and the differing tribunals before which justification may be attempted. Is the call to justify a belief a matter of proving the belief true? Or a matter of showing that it was not foolish to believe it, given the state of knowledge in related fields at the time? Or a matter of showing that the belief was morally acceptable or perhaps merely conventionally acceptable? What has to be satisfied in a justification—an objective standard of proof? An objective test of truth? A reliable human judge? A committee of reasonable people? Popular opinion?

There is no doubt that scientific knowledge has been paraded as a body of provably true information. It has been claimed for beliefs in scientific matters that they could satisfy the toughest tribunal of all: They could be proved so that every rational person sufficiently intelligent and well-informed to follow the argument would be bound to agree, on pain of forfeiture of rational integrity. When I have been discussing full rational justification in this paper, this is what I have had in mind. The upshot of my argument so far is that scientific knowledge does not enjoy this status. We might ask about some weaker sense of 'justify.' Is there not some sense in which we can say that a scientist's belief in a theory now thought false was justified, given the evidence at his disposal? If the point of doing science is to find out the truth about the structure and dynamics of the real world, it is beside the point whether a particular scientist was justified (in some weaker sense) in believing what he did. Of course, it is not beside the point for a biographer delving into the wisdom or folly of a person's belief, to ask questions about the justification of a belief by this or that standard. Very interesting questions, for example, may be raised about Galileo's beliefs. It is now thought, the magnificence of Galileo's contribution to the progress of science notwithstanding, that much of what he believed was false and unsupported by available evidence. One can ask about the theological justification, the justification by agreement with the working scientific community, or within society more generally. One can ask about the moral justification of believing what was so widely rejected when the arguments in its favour were not overwhelmingly convincing, and so on. Nonetheless, the question of the pursuit of truth is hardly answered. If full rational justification is not available, all other justifications seem beside the point. The question to ask in the pursuit of truth should not be, 'How far am I

27

justified in thinking this theory true?', but rather, 'In what way, supposing this theory not to be true, may I set about improving it?' Weighing arguments in favour of a theory may be relevant in the quest of improving it: in deciding whether, for example, to treat it as already improved upon by some rival which should now be the object of critical attention. But such an exercise would be far removed from justifying the theory. The distinction is a matter of intention. The myth of the objectivity of science, which has included the false doctrine that scientific knowledge was fully rationally justifiable, has obscured the important role of scientists' intentions in the growth of knowledge in science. The strategy of individual scientists will differ according to whether their intention is to prove a theory or to improve upon it. Thus, the identification of reasoning in favour of a theory with justifying it is an obstacle to the adoption of strategies aimed at the growth of knowledge.

The distinction between technology and science can perhaps be most sharply drawn on this very question of justification. Even if we assume that scientists need never justify a single theory, it could be argued that technologists require that scientific theories be justified before they design structures or processes which presuppose them. This is a moot point, since historically there was considerable progress in engineering before the development of theories to explain the success, and it is still true in some branches of engineering, though not in all, that technological success runs ahead of theory. Nonetheless, when we focus on applied technology or on the implementation of some novelty, there is undoubtedly a question of justification; the government wants assurances that there will not be too deleterious an effect on the environment or on the social fabric. Thus, as Agassi has suggested, the justification of scientific theories is called for in the context of the implementation of a novelty that applies them practically, rather than in the pursuit of the growth of knowledge.

If we set aside justification as the aim of argument, we might want to ask what character rationality takes on and whether science is still to be considered rational. Actually in recent philosophy of science things went in a different order. The first full attempt to characterize rationality in a non-justificationary manner (by Bastley) developed ideas Popper had after he had given the problem of the growth of knowledge in science a non-justificationary solution. But the order is not important. Let us agree with the common judgment of modern times that scientific inquiry is a species of rational inquiry. The feature which now distinguishes rationality after the rejection of justification as an aim from rationality is openness to criticism. Just as with justification, openness to criticism has a psychological or intentional

28

component. Instead of rationality being thought of as demanding the satisfaction of a necessary and a sufficient condition for the rational tenability of a belief—all and only those beliefs are rationally tenable which are rationally justifiable—under this different characterization it is thought to demand the satisfaction of only a necessary condition for the manner in which beliefs are to be held, namely open to criticism. Rational people will still weigh arguments for and against a theory: They may suspend judgment, or they may assert the theory. In the absence of full rational justification, rationality requires of them that they do not regard their judgment as final.

It is worth remarking that holding a theory open to criticism is not to be confused with doubting it. Nor does holding theories open to criticism rule out acting on them. Thus, the story of the skeptic who was rendered inactive because he did not think any action justified is a caricature. Its subject is a proper object of pity: He thought there were no justifications but had not found out how to do without them, namely to do what you think best without needing the assurance that posterity will be rationally bound to agree with you.

Of course, holding one's theories open to criticism is not a sufficient condition for the growth of knowledge. In addition there needs to be imaginative conjecture of possible improvements over the theory currently thought best, and there needs to be an actual critical appraisal, including stringent tests, if these can be devised. But openness to criticism is a necessary condition for the growth of knowledge. Interestingly enough, it seems to suffice for the community of scientists to be open to criticism; it does not seem necessary for every scientist to be openminded. There is room in science for some dogmatists, even though growth would be threatened if there was a dogmatic consensus.

I should like to draw attention now to just two of the many consequences of adopting a skeptical rather than a justificationist view of rationality. I am not attempting here to justify a non-justificationist theory of rationality—that would be a curious intention—but rather to see the view I hold within the context of the discipline of philosophy of science by drawing attention both to the problem-solving power of my view and to the different selection of problems for study, which it suggests. The first consequence is that growth of knowledge in science, as contrasted with other disciplines, is explained in a more satisfactory manner. The second is that the importance of a particular set of problems is diminished, while a different set gains in importance.

Let us assume that philosophers of science are constrained as to the theory of rationality they adopt as follows: to pursue knowledge by scientific method is assumed rational, hence the theory of rationality adopted has to give an account of scientific method which explains the growth of knowledge in science. A justificationist theory of rationality easily explains growth of knowledge in science if this growth takes place by accumulation. Scientific knowledge will then be that body of sentences (particular and general) which have been rationally justified. Revolutions in scientific knowledge are much harder to explain, since they can take place only if theories previously thought justified rationally are later thought not justified rationally. In the history of science, during most of which a justificationist approach to science has been dominant, revolutions in scientific knowledge have been followed or sometimes accompanied by revolutions in theories of justification— but there is not room here to tell that interesting story. A great deal of attention has been given in twentieth-century philosophy of science, especially in the last few decades, to explaining revolutionary growth of knowledge in science, following the remarkable theoretical revolutions in physics from 1895 to 1930. The pioneer work in this connection was done in 1935 by Popper, whose non-justificationist approach makes it possible for him to explain growth of knowledge. In Popper's theory of science, knowledge grows by conjecture and criticism. All scientific knowledge is conjectural, and the point of observation in science (apart from supplying the scientist with his problem) is to test the conjectural theories to see where they are in error and in need of improvement. Obviously, revolutions in science are more easily explained on this view, than on a justificationist view, which does not even improve on already entrenched theories as an aim in science. Nonetheless, it *is* an aim among scientists and needs explaining.

Secondly, provided that one assumes that rationality requires holding all and only those beliefs which can be justified rationally, and provided that how a belief is justified rationally is not fully agreed upon, the main thrust of philosophical inquiry will be directed towards developing a theory of full rational justification or, failing that, a satisfactory theory of partial justification. Thus, we may explain much of the interest of philosophers of science in solving the problem of induction. When it was agreed that that could not be solved, the problem of the preferability of one unjustified theory over another was substituted. Various candidates were scouted in the hope of finding a single characteristic such that if a theory possessed this characteristic to a greater degree than a rival, it was to be preferred. There was not general agreement as to what that property should be or, given a variety of properties, why any should be preferred over another, but the debate has centred on a few such characteristics of which explanatory power and degree of confirmation have perhaps been the most important. Popper was the first philosopher in modern times to adopt

30

a deliberate and explicit non-justificationist theory of knowledge. His work has opened up a new range of problems, aside from ongoing problems of giving an account of rationality and using that to explain how human beings, thought rational, can indulge in magic, religion, and the like. For example, a new approach to metaphysics is possible. Inquiry in metaphysics has usually been prosecuted a priori, and metaphysical findings have been offered dogmatically, their justification thought secure in some a priori principles. If the rejection of justification is carried over into metaphysics, and if, furthermore, the attempt is made to hold theories in metaphysics open to criticisms rooted in science, some interesting problems become tractable, as Agassi, Bunge, and I have found. Philosophical attention is shifted, given Popper's approach, away from the problem of showing how scientific knowledge may be justified (with attention focussed on logic and methodology), towards a criticism of the findings of science. Whereas from the justificationist point of view, philosophy seems irrelevant to science, since it does not aim either to add or to subtract from the body of scientific knowledge (that being the job of scientists), from a skeptical point of view, it is part of the philosophical enterprise to criticize scientific theories, to challenge presuppositions, and, more generally, to take part in the scientific enterprise of describing reality and explaining experience. Incidentally, one of the important problems for the positivists in this century has been to demarcate science from non-science, especially from religion and metaphysics. No adequate solution has been forthcoming. Given a skeptical theory of knowledge, this problem sinks in importance, since ideas or theories from religion and metaphysics not held dogmatically can be criticized and the resulting debate may advance our understanding of reality. The problem of distinguishing science from non-science or one branch of science from another remains, of course, but its importance is largely political: If we cannot tell the difference, we do not know to which funding agency to apply for a research grant.

6. The Rational Component in Justification

Earlier I promised to return to the matter of scientific knowledge being more than a set of sentences. That is as good a way as any to begin discussion of the second main alternative to the failed positivist critique of science, the alternative that yields to the authority of scientists to supplement reason, since its earliest exponent was Polanyi, who made much of the tacit knowledge of a connoisseur in explaining why he thought the nearest we can get to truth is the passionate asseveration of an expert. I am very sympathetic to Polanyi's intention

31

to play up the importance of inexpressible knowledge in giving an account of the judgments people make. The positivists' account of science certainly ignored it. Not paying attention to the difference between what is reported and what is perceived allowed the positivists to play down a number of important problems which, as I have already suggested, undermine their programme. Thus, I prefer Polanyi's view of personal knowledge to the earlier view. In my view, each one of us brings a tacit component of knowledge to bear in the comprehension of any sentence in science, a component informed not merely by the overrichness of our perceptual field, but also by metaphysical speculation. Nonetheless, I do not agree with Polanyi's suggestion to defer to the expert, and this not just because I think individuals may be wrong. I reject also Ziman's attempt to improve on Polanyi's *Personal Knowledge* with his *Public Knowledge* in which scientific knowledge is identified with the consensus of the scientific community. I reject both views for much the same reasons, one of which is that while they explain why scientists know more about science than ordinary people they do not explain why I should trust them. Both Polanyi and Ziman agree with me that full formal rational justification is not available. They agree that scientific knowledge is subject to change. Nonetheless, they claim my trust. Perhaps they do not think I should trust scientists in everything; but they give me no criterion by which to detect when to trust and when not. Ziman, in trying to overcome the problem of individual's errors, raises another: How is the layman to detect when the scientific community has reached a consensus on an issue? This problem is especially acute in matters of life and death or at least of comfort and discomfort. In some disciplines there are schools of thought: There is unanimity within the schools, but division between them. What is a layman to do then?

I believe that there are good reasons why people should often trust scientific opinion, and these also serve as a guide to the layman in choosing when not to follow scientific opinion. But they come from a non-justificationary theory of rationality and of science. I shall come to these in a moment. Let me first give an additional reason for rejecting authority as a supplement to reason in the justification of a belief or a course of action: It offends my moral autonomy by closing off the option of further criticism and removing from me the responsibility for my own beliefs. I have no objection to a theory that says that responsible belief requires serious attention to the opinions of experts, and I do not object to the constraint placed upon my freedom of action by my belonging to social institutions which may have rules that could silence my public criticism of some belief or action (a rule for closing debate in a public meeting, say). But I reject a theory of belief that takes away my freedom to decide for myself what to think by announcing that connoisseurship, consensus of experts, or some such decided matters of truth for me.

The opinion of scientists is often to be trusted mainly because they have already critically examined most of the alternatives and have found problems with the ones they have rejected. A quick test to show whether to trust a scientist's opinion on a particular issue is to ask what alternatives have been criticized and rejected. (One could always give one's own pet theory a try by inviting criticism of it.) The issues on which scientific opinion is not to be trusted are those where the opinion is based on presuppositions one rejects or where an inadequate survey of alternatives has been made—this regardless of any unanimity. Unaminity among experts in a small field is too easy to explain on other grounds than that they are right for one to be impressed by unanimity itself. Scientific opinion is impressive when it resembles the skeptical model as the result of serious attempts to eliminate errors, and thus holds out some promise of avoiding the penalties the most obvious errors might have carried.

Interestingly, the same kind of consideration weighs in deciding whether to trust decisions of courts of law. It is not the expertise of the lawyers and judges that gives promise that a decision on guilt or innocence will be correct, but rather that the system requires a decision of guilt to be beyond reasonable doubt. Reasonable doubt is removed when all the alternatives ingenious minds can conjure have been weighed and found wanting.

I have written above of varieties of justification. A few loose ends could usefully be tied by returning to those varieties, of which three are worth review: rational, moral and social. Moral and social justification could be played down in the context of the doctrine that true beliefs could be given thorough rational justification, especially if correct action could be subsumed under true belief. Society could hardly properly gainsay action based on true belief. Yet, some societies did persecute and even execute persons we now think to have been right and proper in thought and action (at least respecting the particulars of their "offence"). Where this happened, social justification was impugned. Things are different in the context of a doctrine that denies that final rational justification can be given, coupled with a doctrine of the moral autonomy of human beings. Moral considerations related to the risk to others consequent upon one's being wrong may now be entered. Given differences of opinion among individuals, a social institution for arbitrating differences will take the stage. People wishing to undertake activities dangerous to others will need a licence to do so. An action is socially justified, relative to a particular society, if the action is licenced or permitted by the society. In a sophisticated society, scientific opinion will be weighed in relevant matters in deciding what to permit and what to prohibit. Science has much to contribute to government that is of benefit, provided the contribution is in the context of critical debate. A rational component thus enters social justification.

Moral justification is a matter of satisfying moral standards or considerations. It is possible for there to be a rational component in such justification also, if it is done with openness to criticism. Even so, there cannot be moral justification of the *standards* of morals. If there is an objective moral obligation, morality cannot justify it. Nor can argument. But objective moral standards are no more beyond the reach of reason than are theoretical truths in physics, for example, though they may be harder to agree on. A skeptical rationality calls for the proposal of conjectured standards which must be open to criticism.

Similarily with social justification: An action may be justified according to a society's standards, but the standards themselves can hardly be justified by social decision—though they are certainly adopted, or not, according to social decision. Standards of social justification can be criticized both rationally and on moral grounds. There may even be moral justification, which would, of course, itself be subject to criticism. No justification quite eliminates personal responsibility for action, though, as I have already suggested regarding technology, social justification represents a sharing of responsibility.

To summarize: If knowledge in science is not intended to lead to action, justification is beside the point—which is a relief seeing that full rational justification is not possible. But if knowledge is to be used as a basis for action, questions of moral and social justification arise. The rational component both in the growth of scientific knowledge and in decisions about morality and social approbation is a matter of weighing alternatives critically and choosing responsibly, rather than a matter of justification.

Bibliography

Agassi, J., "The nature of scientific problems and their roots in metaphysics," in M. Bunge (ed.) *The Critical Approach to Science and Philosophy*, Free Press, London, 1964.

Agassi, J., "The confusion between science and technology in standard philosophies of science," *Technology and Culture*, 7, (1966), pp. 348–366.

Agassi, J., *Science in Flux*, Reidel, Boston, 1975.

Bacon, Sir F., *Novum Organum*, 1620.

Bartley, W.W., *The Retreat to Commitment*, Knopf, New York, 1962.

Bunge, M., *Method, Model, and Matter*, Reidel, Boston, 1974.

Carnap, R., *Logical Foundations of Probability*, Chicago University Press, Chicago, 1950.

Descartes, R., *Discourse on Method*, 1637.

Hanson, R.N., *Patterns of Discovery*, Cambridge University Press, Cambridge, 1965.

Hempel, C., *Aspects of Scientific Explanation*, Free Press, New York, 1965.

Hume, D., *Inquiry Concerning Human Understanding*, 1748.

Kant, I., *Critique of Pure Reason*, 1781.

Kant, I., *Prolegomena*, 1783.

Kuhn, T., *The Structure of Scientific Revolutions*, Chicago University Press, 1962 (Second enlarged edition, 1970).

Maimonides, M., *Guide for the Perplexed*, 1194.

Nagel, E., *The Structure of Science*, Harcourt, Brace and World, New York, 1961.

Polanyi, M., *Personal Knowledge*, Routledge and Kegan Paul, London, 1958.

Popper, Sir K.R., *The Logic of Scientific Discovery*, Hutchinson, London, 1959 (translation of *Logik der Forschung*, Vienna, 1935).

Popper, Sir K.R., *The Open Society and Its Enemies*, Routledge and Kegan Paul, London, 1945 (4th edition, 1962).

Popper, Sir K.R., *Conjectures and Refutations*, Routledge and Kegan Paul, London, 1963.

Robinson, H.J., *Renascent Rationalism*, Macmillan, Toronto, 1975.

Settle, T.W., "The Rationality of Science versus the Rationality of Magic," *Philosophy of the Social Sciences*, I, 1971, pp. 173–194.

Settle, T.W., "The Relevance of Philosophy to Physics," in M. Bunge (ed.), *Problems in the Foundations of Physics*, Springer-Verlag, New York, (1971) pp. 145–162.

Settle, T.W., "Induction and Probability Unfused," *Philosophy of Karl R. Popper*, P.A. Schilpp, (ed.), Library of Living Philosophers, (1974) pp. 697–749.

Settle, T.W., *In Search of a Third Way*, McClelland and Stewart, Toronto, 1976.

Settle, T.W., Jarvie, I.C., and Agassi, J., "Towards a Theory of Openness to Criticism," *Philosophy of the Social Sciences*, **4**, (1974) pp. 83–90.

Ziman, J., *Public Knowledge*, Cambridge University Press, Cambridge, 1968.

The Hypothetico-Deductive Model of Scientific Theories: A Sympathetic Disclaimer

ROBERT E. BUTTS

The major philosophical points I wish to make in this essay are not fully original, although I hope that what I have to say makes a coherent and persuasive whole. I shall be discussing a predominating philosophy of science that has held the field against most other systematic philosophies of science for at least five decades. Sometimes this philosophy has been called "Neo-Positivism," (in part to distinguish it from the nineteenth-century positivism of Comte) sometimes, "Logical Empiricism." It shares much with the early "Logical Positivism" of Schlick and other members of the Vienna Circle—a sophisticated philosophical position not to be confused with the positivism of Ayer's shallow *Language, Truth and Logic*—but for many advocates the historical roots lie elsewhere. The labels are not that important, especially since every essential ingredient of the philosophy of science I shall be discussing is to be found in much earlier, diverse philosophies. It is important, however, to pause and consider this historical situation. There are many twentieth-century philosophers who think that philosophy is an autonomous discipline whose history can be ignored; some contemporary philosophers regard philosophy of science as having come into existence only in this century, or maybe in the late nineteenth century. In a typically misleading article on the history of philosophy of science in the prestigious *Encyclopedia of Philosophy*, P. Harré begins philosophy of science in England with the nineteenth-century debate between William Whewell and J.S. Mill on the nature of induction, claiming that it was only then that issues in philosophy of science and general issues in theory of knowledge were separated from one another. This claim totally ignores that for centuries science *was* philosophy, and also that at no time have epistemological issues been independent of those special questions now claimed to lie within the domain of systematic philosophy of science. Aristotle's attempt to account for causal and demonstrative explanations in physics and biology is surely as relevant to concerns of philosophers of science as is Mill's *Logic*, indeed, probably more so. Perhaps, then, I will be forgiven if I take a somewhat more sweeping and charitable attitude toward both philosophy of science and its genuine historical embodiments.

Developing a philosophical characterization of the nature of

36

science has never been an easy task, partly because of the range and complexity of the problems involved and also partly because the sciences have developed so rapidly that study of relatively stable parts of theories has been impeded. Nevertheless, the history of philosophies of science (at least that section of the history that extends from Galileo to recent logical empiricists) does seem to be represented by a series of views that converge in the twentieth century on a picture of science that might be called the standard hypothetico-deductive model of scientific theory. Any particular philosopher of science selected for discussion would probably differ from some other philosopher on matters of substance both in detail and in programme. If we allow for such programmatic and specialized differences, however, I think that the main features of the hypothetico-deductive model of scientific theories would include at least nine essential ingredients. After developing the model I will attempt to isolate some of its fatal faults.

To begin: On this model a scientific theory is (1) a partially connected set of sentences of different degrees of logical generality. On the basis of this condition, one might refer to the theory as the *sentential* view of scientific theories (which would include, of course, that the *equations* employed in the various sciences be construed as sentences). The view is perfectly straightforward; it is also controversial. As a set of sentences each well-articulated science will contain some generalizations, some empirical claims asserted to hold for all individuals, situations, or events of a certain kind. Some of these universal generalizations will be regarded as *laws*, logically general assertions with counterfactual import, meaning simply that these generalizations will hold (be true) whether or not the individuals or circumstances they talk about continue to be observed. Put differently, science is here regarded as telling us what would happen *if*, and we want to regard its laws as true even if the *if* is seldom actually realized.

Not all universal generalizations will qualify as law-candidates. Thus, "all automobiles have four wheels" needs to be distinguished from "all mammals are warm-blooded." The statement about automobiles is referred to as an *accidental* generalization; if the sentence is true, it is true only because of an accident of history. Most of us would be quite satisfied to call something an automobile if it had fewer or more than four wheels. On the other hand, the statement about mammals makes a stronger assertion. The assertion is not only true (we do observe warm blood in all mammals), but we would want to continue to think that it is true if the species mammalia were selected out of existence. And it would still be true that *if* something came along that was a mammal it would be warm-blooded. It should be noted, however, that although the distinction between accidental and lawlike generalizations is intuitively clear and surely desirable, philoso-

37

phers have had a notoriously hard time developing the logic of the distinction. (Involved in the question of the counterfactual character of scientific laws is the question of *scientific realism*, a matter of some importance to which I will return later.)

In addition to laws (or lawlike contentions), each such sentential science will contain generalizations of a more impoverished kind, sentences of the form "Under the following very determinate and normally repeatable circumstances, the following event(s) will occur." Sometimes these sentences will be classed as experimental laws or empirical generalizations to contrast them with fully lawlike sentences. They are totally contingent logically, and hence any might turn out to be false. Nevertheless, such generalizations are normally indispensable in specifying the initial conditions under which a fully lawlike generalization can be instantiated and hence subjected to test. At the bottom rung of the ladder each H-D (adopting this rubric as shorthand for hypothetico-deductive) science will contain existential generalizations and statements about actual spatio-temporally fixed events, for example, sentences of the form "There are objects with such-and-such properties," and "At time t event e took place." These sentences, suitably interpreted for each science, will furnish the principal means for connecting the theory to the "facts" or "data."

(2) The principal machinery employed in stringing together the sentences in the kind of theoretical framework just outlined is the logic of deduction. In the ideal case (rarely, if ever achieved), the theory is fully axiomatized, and each theorem about specific individuals or events will be a logical consequence of the axioms under some suitable empirical interpretation. More usually, the theory will not be fully axiomatized, but an attempt will be made to get as much deductive or calculational strength into it as possible. I said above that an H-D theory is a partially connected set of sentences. I can now be more specific about this caveat: Deductive power is the goal of science; the problem is that a strict deductive connection is not often found to obtain between important sentences in a given science, and if it is found to obtain it is also found to be circumscribed by a set of background sentences (sentences stating initial conditions that instantiate the antecedent of conditional generalizations) that render it somewhat weaker than is normally desired in cases of formally logical deduction. This is true even in those cases where the science is represented mathematically. Very few mathematical "proofs" possess the clean lines of pure deduction. One can understand the initial plausibility associated with wanting scientific systems to have the kind of logical integrity that deductive systems possess. The history of science provides plenty of examples of attempts to get a conceptual

picture mirroring a "unified" structure of the world: Newton's inverse square law as a unification of laws apparently so unrelated as Galileo's acceleration-rate law and Kepler's laws of planetary motion; Einstein's attempted unification of relativity theory and quantum theory in a more general "field" theory. The lure of deductive structure is intimately connected with this epistemological ideal of systematic unification of theories. It is also associated with what one might call *pragmatic* dimensions of science, most especially with parsimony principles used as extra-evidential measures of the acceptability of theories—in substantive form: "Nature does nothing in vain"; in evaluative form: "Seek methodological simplicity." I will return to certain additional features of the sentential view of scientific theory that require employment of the deducibility condition.

In suggesting that a sentential theory contain at least some generalizations which might be laws, I was hinting at what is a most important requirement of such theories: (3) In order not to be regarded as merely empty logical formalisms, they must be "about" something, must be true or false of some "reality" which exists independent of the theory. Seeking to avoid what they regarded as the unfortunate consequences of so-called "subjective idealism" many philosophers have tried to introduce a justified extra-mental or extra-theoretical subject matter that would tell us the way the world is, even if that world should cease to contain knowers. The history of philosophy is full of attempts to establish this external reality. Some thought that we could have direct intellectual intuition of such a reality; others limited their reality to what could be achieved in, or reduced to, the actual having of direct sensations. Kant, thinking that he had discovered epistemological excesses in both of these views, articulated the idea that there is no knowledge without both form (theory) *and* content (that which is given independent of theory), thus creating a theory that is in some respects a prototype of the sentential theory we are investigating.

Later I shall discuss some of the notorious and much-argued difficulties surrounding this concept of the "given." At this point, however, I want to suggest some reasons for the initial plausibility of capturing this elusive philosophical creature. Anyone who pretends to scientific *realism*, anyone who holds that his theories must face the tribunal of independent "facts," will need to think hard about his concept of the given. In the way that C.I. Lewis put it, the idea is quite simple by comparison with more grandiose philosophical ideas: The given is that which we cannot, by taking thought, alter. No amount of thinking will alter *that* I have a pain in my right foot; no amount of thinking will alter *that* I crash into my study wall and am repelled by it if I try to walk through it. Now Carnap and other advocates of the sentential theory sought to capture this feature of "hard" experience by working with the rather more sophisticated concept of an *observation language*, a language that in some way directly reports observationally inspectable features of the world. In the sciences, in physics for example, this observation language will specify the operational or observational meanings of the terms required in order to *apply* the theories, terms like "straight line," "length," and the whole array of physical and geometrical terms introduced by the theories. Of course, not all of the terms must have such an observational specification of meaning. Physics will continue to contain some purely theoretical terms; it is only required that some of the terms in the theory be linked to observational situations by means of the observation language. If enough of the predictions of a theory thus observationally linked with experience turn out to be correct, then the theoretical terms which are not directly mapped onto experience will be epistemologically justified. We are entitled to think that there are electrons, not because we directly encounter them in our experience, but rather because theories which talk about them turn out to have many correct predictions. (Thus the sentential theory of theories contains an implicit ontological criterion: Something exists—even if that something is only hypothetically postulated—if there are theoretical terms describing it in a theory whose linkage with observation results in correct predictions.)

It is clear that if condition (3) is to be met a further condition must also be satisfied. Condition (4) follows from what I have just been saying: If a theory is to be about something other than itself, the theory cannot be the sole arbiter of meanings of terms. The meanings of terms employed in the observation language must be given independently of the meanings of the theoretical terms as specified in the science. If we can somehow locate an independent given then it will so specify these meanings. If this quest fails, then the best we can do is try various observational and experimental specifications of terms, and then apply the theory under such conditions. At this point the apparently clear distinction between a theory and an observation language begins to become quite murky. For if we are to try various experimental languages, then either we shall just *say* what the terms will mean, or we shall attempt to generate a set of *experimental laws* which will be different from, and have different legal credentials than, the theoretical laws. For example, to learn how to manipulate nature so that we can observe how far objects of differing weights fall in a vacuum is to collect a number of "facts" that give us merely empirical generalizations over the experimental contrivances. But what is the relationship between these experimental "laws" and Galileo's Law, a theoretical law

40

that tells us that in a state of free fall an object moving in the direction of the Earth accelerates at the rate of 32 ft. per second[2]? Clearly in order to connect the experimental laws (remember that the meaning of their terms is given without recourse to Galileo's theory) to Galileo's Law we will need an additional law or laws telling us how to "translate" the theoretical statement into the experimental one. Some philosophers have introduced what they call "bridge laws," and "semantical rules" in order to solve this problem. But here I want simply to observe that the problem exists and is a consequence of a seemingly harmless demand that theories be about some "world." In addition, it is interesting to remark that philosophical theories about theories (in this case scientific theories) have the remarkable feature that given sufficient articulation, the philosophical theories begin to generate internal problems interesting in themselves, but that these internal problems take us farther and farther away from our initial task of understanding *something else*.

As serious as this problem of bridge laws may be for some H-D theorists, I am not sure that their account requires this extent of complication. The ontological criterion of the H-D theory allows us to circumvent this problem in the following interesting way: If Galileo's acceleration-rate law is properly formulated in the calculus, then the number of feet that a body falls in so much time (in a vacuum) is a strict *deductive* consequence of the mathematically-formulated law. True, we must stipulate what it means observationally to measure distances and times, but such semantical tampering is relatively harmless, provided only that we have some means of experimentally approximating conditions in a vacuum. Given that this can be done, then any correct prediction of the law will tend to strengthen our confidence that it is right, and each such increase in confidence will bestow some ontological assurance that there are such things as acceleration rates—even though acceleration rate is a theoretical construct to which we have no direct means of access by empirical means. At variance here is the H-D ontological criterion and the apparently harmless wish to have theories *directly* tested by means of observation-language claims. Saving the ontological criterion involves in some sense making observation depend upon theory; saving direct observationality involves leaving theoretical terms in some epistemological jeopardy. There is a dilemma here that I shall exploit in my critical remarks about the sentential view of theories.

If, as the above account of the H-D theory entails, each science will have to be linked to a relevant observation language, then it follows (5) that any theories that compete against each other must do so in the same domain of possible facts. And it follows from this

41

condition on competition between scientific theories that each competing theory will contain some terms whose meaning is not at variance with terms in the theory against which it competes: In a given science any allowable theory will have to be so constructed that it contains at least some sentences that imply sentences of the standard observation language "governing" the domain of facts already in some way handled by the earlier theory or theories. Competition between theories—classical mechanics and relativistic mechanics, for example—is thus always to be seen as an attempt to gain explanatory mastery over the same, independently interpreted domain of facts. Theories can change; what theories are about remains constant. For theories to be "about the same thing," they must contain terms which yield construction of at least some sentences in each science which imply common observation sentences in the preferred observation language.

There can be no doubt that characteristic (5) is a consistent consequence of (3) and (4) and that any good H-D theorist must accept it. But the price of this acceptance is high. There are now additional burdens on common observationality and the given. Whatever special problems arose from problems of maintaining conditions (3) and (4), (5), though logically implied by (3) and (4), now places on science a methodological condition difficult, if not impossible, to realize. Many cases are easy: We can perhaps reconstruct classical physics so that both the wave theory of light and the particle theory of light are competitors in the required sense. But an interesting clash between global physical theories like those of Newton and Einstein is difficult to reconstruct in the conditions set by (5). To oversimplify: Relativistic mechanics is not just asking us to replace the old and accepted mechanics of the world of moving objects with a new theory of the way those "objects" move, it is in some sense telling us that the world of moving objects is quite different from the way we thought it was. The conceptual change we are invited to make is just too large to be put off as a competition between two theories for the same ground; it is more like being told that acceptance of the new theory cuts the old ground out from under us (and replaces it with a much less substantial stuff!). Again an obvious ingredient in the H-D theory places it in an unenviably weak position—and the props may by now be in short supply.

Characteristic (6) of the sentential theory of theories has to do with its commitment to a certain structure of explanation, one that is ancient, honoured, plausible, and difficult to expell. Not all positivistic philosophers of science believe that it is the business of science to give explanations. I am here, please bear in mind, giving the most "liberal" reading of the H-D model that I can imagine. And on this model there is agreement with non-H-D theories that perhaps the most central feature of science is that it tries to be explanatory, it tries to tell us what will happen when and if, to tell us what causes what. The most rigorous attempt to expound the conditions of explanation in an H-D context is found in what is now called the "Hempel-Oppenheim covering law model of explanation."

In their classic paper, Hempel and Oppenheim construe explanations as involving a special relationship between sentences of certain kinds. And this identification of explanation with certain logical connections between sentences is, of course, entirely in keeping with the kind of methodology for science that we are investigating. They consider an explanation scheme to involve two parts, an explanandum and an explanans. An explanandum is an empirical sentence describing what it is that requires explanation. To refer to this sentence as "empirical" already prejudices the model, but we will see shortly that the prejudice is modest and easily agreed upon. The explanans is a set of sentences pretending to account for that which the explanandum describes. The question is: Under which conditions can the explanans be said to give such an account? What are the conditions for an adequate explanation? According to advocates of the covering law model of explanation, a proper explanation must satisfy three logical conditions and one empirical condition. The logical conditions are straightforward: First, the explanandum must be a logical consequence of the explanans; the requirement is strict in that one must actually be able to show that the explanandum is deducible from the explanans. (Notice again how easily the idea of deductive connection infiltrates the sentential model of theories.) Second, the explanans must contain at least one general law. Third, at least some sentences in the explanans must have empirical content: In some relevant and specifiable way, the explanans must have something to say about the "world" as we experience it. Finally, a good explanation must meet the empirical condition that the sentences constituting the explanans must be true.

These conditions for an adequate explanation are controversial. Thousands of words have been written in criticism of the model; thousands more in an attempt to vindicate it. Like most philosophers of science working today, I do not find the model totally satisfactory and shall say why below. As with the other features of the H-D theory, however, I think it only fair to point out the strong characteristics of the covering law model, and to stress its initial reasonableness. It has long been thought that the central business of science was to produce laws which explain various natural phenomena. It makes no difference *why* one wants to find these laws. One might be interested in achieving maximal control over nature or other people for reasons good or ill, or one might simply want to know what is lawful about nature. But no such utilities can be attained without knowledge of laws, from which it seems to follow easily that laws do indeed have this centrality in scientific explanation. The Hempel-Oppenheim model tries to capture at least this intuition about the nature of science. So it appears harmless to require that a proper explanation must contain at least one general law, and that this law should be about empirically ascertainable features of the world. Furthermore, it again appears quite harmless to require that the law(s) be true, for it complies with our intuitions about good explanations that what does the explaining (the explanans) not be false or questionable. (There is a problem here: Many are prepared to accept an explanation tentatively if the explanans is only *probable*—quite likely to be true on the basis of *available* evidence. But one must be careful to distinguish between accepting an explanation as true, and accepting an explanation as true until further notice.)

The empirical condition that the sentences constituting the explanans must be true, and the logical condition that the explanans must have empirical content are true for reasons following directly from other features of the H-D model. It is logical that any false sentence logically implies every sentence; it is also logical that no sentence lacking empirical content can imply any sentences having empirical content. This is only a problem, of course, if we insist upon the deducibility condition of the model, but then if we were to abandon it, there is no longer any clear sense in which a general law "explains" or subsumes statements about aspects of the behaviour of the world. At least the following benefits accrue from retention of the deducibility condition: Armchair scientists will be prevented from offering as acceptable explanations whose explaining sentences are either false or highly unlikely; and explanations without empirical content will place a heavy burden on their advocates to show us how the empirical world of ordinary experience is related to that which somehow transcends the observable world altogether. It may at once be suggested that the sentential theory of theories thus places "metaphysical" explanations at a disadvantage. Many adherents of this theory would insist that we must find deliberate ways in which this disadvantage be made into an advantage for philosophy of science—by simply abandoning any but scientific explanation. Unfortunately, this seemingly neat move is far from conclusive. One of the major anxieties of H-D theorists in this century has been the problem of banishing metaphysics.

Considerations of this kind lead us naturally to property (7) of the sentential theory of science: Each proper scientific theory will have to be *testable*. This property of scientific theories has been prominent, although not specifically noted, in much of our previous discussion—certainly conditions (3) and (4) have several important points of contact with issues of testability. Here also the theoretical demand appears quite uncontroversial: Everyone knows that it is just this feature of science that sets it apart from other cultural enterprises. The claims of science, so the familiar story goes, can be checked against what we observe either directly or as the result of experimentation. Again, the "facts" determine the truth or falsity of the claims of science. The distressing thing is that scientists and philosophers of science have experienced enormous difficulties in working out a myriad of problems surrounding the concept of a test. It is true that at a certain simple level at which what is at issue is easily agreed upon the results of checking the facts can be relatively decisive. I can fairly easily determine if the Australian rubber tree in my study is sending out a new shoot by entering the study and observing the plant. Of course, we must make modest assumptions about the intelligence of the observer, his adequate eyesight, and other factors. But not much turns on the outcome of this slight challenge, and we are thus prepared to let the "facts" tell us the way it is. But unfortunately the situation in actual science is just not simple in any of these ways.

Certainly testability is not a straightforward matter when it is viewed as a requirement of the sentential theory of theories. As we have seen, a sentential theory is not directly testable by confrontation with experience—many of the terms in such a theory are hypothetical and have no direct empirical counterparts, not even in principle. It is the observation sentences entailed logically by a theory which are subjected to direct test. Positive instances of the observation sentences, when discovered, will give credence to the theory which entails them, but no amount of positive evidence will finally establish a theory as true once and for all. The reason for this is simple and logical: Our scheme is of the form, "If the theory is true, then the following observation sentences will also be true." But we cannot know directly that the theory is true, we can only know that it is true via the truth of its observational consequences. In logic, the inference pattern: If p, then q, q, therefore p, is not universally valid, which means that no amount of confirmation of q will forever establish the truth of p. No matter how strong the evidence is for q, p might still turn out to be false. To use our earlier example: Suppose p to be "All mammals are warm-blooded." q will actually be a set of q's of the form "a is a mammal and is warm-blooded," b is a mammal and is warm-blooded," etc. Each member of the set of q's will be an instance of p and will confirm p, but no number of q's will establish the truth of p just because it is a universal generalization which might be false even though all known q's are true. Proceeding in the way in which the H-D model must proceed thus generates large problems: H-D theories can be found to be highly confirmed by the evidence, but they can never be established to be true.

Confronted with this unpalatable logico-epistemological result some philosophers simply settled for the weaker situation: We cannot gain scientific truth, but we can turn our attention to the degree in which it is correct to say that one theory is better confirmed than another. (I shall not here enter into the labyrinthine difficulties that have been discovered in the logic of confirmation.) Others—Sir Karl Popper and his followers are the most vigorous defenders of this view—think we should start our reconstruction of the methodology of science from the premise that if we cannot conclusively establish a theory, logic does permit us conclusively to falsify some theories. On this falsificationist way of proceeding, logic is on our side. If the inference scheme: If p, then q, q, therefore p is invalid, the following important deductive inference scheme (called *modus tollens*) is not: If p, then q, not-q, therefore not-p. A theory T implies its observational consequences E. But if just a single e in the set E turns out not to be the case, then T will be false. It is an important correlate of falsificationist method that one should choose theories that imply a great deal; given large empirical content, they will be easily refuted. Theories with large entailment content which withstand repeated rigorous tests are not shown to be true, they are corroborated in the sense that they have withstood attempts at falsification. There is a certain intuitive merit in falsificationism. The history of science shows fairly decisively that theories once taken to be true are subsequently replaced by other theories. Thus, there is an important historical respect in which science is not getting us closer to the truth by multiplying the number of known-to-be-true theories. On the contrary, history shows inductively that for any given theory at present accepted as true, there are good odds that we will discover later that it is false. Thus the falsificationist view seems closer to the actual movement of the historical epistemology of science by urging us to see the history of science as the history of discarding false theories.

As one might expect, however, things are not all that orderly and joyous in the falsificationist camp. It turns out that whatever problems there are with the logic of confirmation are mirrored in the logic of falsification. Moreover—and this is the only point of logical and epistemological detail that I will introduce—there is a difficulty at the very heart of falsificationism, one first pointed out by Pierre Duhem in his influential nineteenth-century book, *The Aim and Structure of Physical Theory*. The problem is that any powerful theory, like the global theories of physics, is not a *single* generalization counting as a theory, it is rather a *set* of interconnected theories plus the background information needed in order to apply such theories. Hence a well-articulated scientific theory faces empirical testing as a body of hypotheses about the world, not as an isolated single hypothesis. The logic of this situation is important enough, I think, to merit a moment's close attention. We have been thinking of a single theory T tested by its set of observational consequences E. Now *modus tollens* tells us that in the simple case where T implies E and E is false, so also is T false. This is a matter of deductive logic: A conditional statement cannot be true if its consequent is false. If T is, however, a set of conjoined Ts, then the logical situation is more complex and needs to be represented as follows:

$(T_1$ and T_2 and T_3 and $T_n)$ implies E.
But E is false.
Therefore $(T_1$ and T_2 and T_3 and $T_n)$ is false.

But if a conjunction of statements is false, then (in the case being considered) either T_1 or T_2 or T_3 or T_n is false, but logic alone has no way in which it can tell us how to decide what particular T (or perhaps all of them) is false. It seems literally to be a matter of choice, one that can be made on any grounds whatsoever. From the logic of this situation, Duhem concluded that no single physical hypothesis is testable in isolation. This has come to be known as the "contextualist" thesis, or the "Quine-Duhem" thesis (the American philosopher of science and logician W.V.O. Quine having championed a form of the view with much wider application to all of science). The thesis has obviously powerful and, for some philosophers, unwanted consequences. Certainly, any form of the falsificationist theory of scientific method must face the challenge of contextualism. But there are other difficulties engendered by the thesis; one of them has dominated much discussion of the foundations of physics.

Duhem's contextualism raises the possibility that in the case of important global scientific theories we must acknowledge that decision between theories can no longer be regarded as a matter decidable simply by putting those theories to empirical test. Something in addition to the "facts" will have to be appealed to in such decisions, a consequence that gives rise to a number of theories of science that argue that in the absence of hard factual decision procedures certain extra-evidential measures of the acceptability of theories will have to be appealed to. Kemeny, Goodman, and other suggested that *simplicity* of theories will now come to play its long awaited and justified role in assessing theories. For those scientists and philosophers concerned to maintain favoured theoretical constituents against apparently recalcitrant evidence, Duhem's thesis regally provided a way out which has come to be known as "conventionalism." Poincaré and Einstein, both disturbed by the consequences for physics of the new non-Euclidian geometries, appealed to a form of contextualism that puts the challenge of Duhem in sharp relief. Both accepted that no single physical hypothesis is testable in isolation; both drew the conclusion from this that when a physical theory is in empirical trouble, it is the *geometry* that must and can be saved, whatever other parts of the physics must be surrendered in the face of falsifying instances. Both took contextualism as a way of saving Euclidian geometry, but this salvation must be seen as purchased at a very high price. Geometry is critical in the application of physical hypotheses to the world—it provides, when suitably interpreted, a spatial metric by means of which we experimentally measure distances. But if a certain geometry is to have a privileged status in physics even in the face of falsifying cases then the observation language of physics is likewise privileged. This will mean, however—recall here the discussion of condition (4) of the sentential model—that the important spatial terms of the observation language are determined by *theory* (in this case geometry), and hence that the ontology of physics will in large part be fixed, not by successful tests, but by commitment by fiat to a certain theoretical structure. Conventionalism, though forced upon us by Duhem's thesis, leaves us with the stark realization that theory determines observational meaning, that no independent source of the meaning of observation terms is available, and that testing as "consulting the facts " plays a relatively less distinguished role in science than we would have wished.

While much that needs to be said about the H-D model of theories has already been discussed above, two additional features require at least brief mention. The first, condition (8), bespeaks again the extent to which the model appeals to deduction. The H-D theory is committed to *reductionism* in several forms. First, in terms of the model we are required to regard *progression* within a particular science satisfying the model as progression toward greater generality of the laws ingredient in that science. This means that lower-level laws will be "reduced" in the manner of being subsumed under, shown to be implied by, the laws of greater generality. Progression in physics was thought to have been attained when Kepler's laws, Galileo's law, and empirical generalizations describing the behaviour of the tides were all reduced to—thought to be logically derivable from—Newton's inverse-square law. William Whewell discussed this phenomenon of reduction in his nineteenth-century *The Philosophy of the Inductive Sciences*, referring to scientific progression as a "consilience of inductions": a convergence of laws derived from different domains by induction upon a single law capable of explaining that such inductive differences can now all be avoided by seeing that the data were after all reports about things of the "same kind."

The second form of reduction is a simple generalization of the first. It may be that some particular science explains, not only the objects and events native to its theoretical involvement, but also objects and events of a different science altogether. What is here displayed in the H-D model is the time-honoured quest for the "unity of the sciences," a unity presumably built upon the foundations of a single foundational science. For some this foundational science is physics, and the many successful attempts to thus "reduce" other sciences to physics has had spectacular results in this century. Both forms of scientific progress through reduction have been much discussed in detail—by Ernest Nagel, for example. But the fundamental point to keep in mind is that the sentential theory of theories continues the age-old "Chinese box" image of scientific progress, where each container box is more general, more "consilient," than the boxes it contains. The "boxes" are, of course, deductively related, and the final greatest achievement of science would be the discovery of the science of all sciences, the generalization that would unify all previous generalizations. Once more it can be said that the writings of Whewell display this image in what is probably its clearest form. One continues to wonder: If deductive structure is weak and wanting, and if direct observationality is plagued by the illnesses I have diagnosed, how, without appeal to convention, can one assess the claims of a pretended reduction, a portended unification? How, that is, within the limits set by the sentential theory of theories?

Lurking in the wings as we have moved along in this discussion of the H-D model—and even muttering an occasional aside—is what I shall arbitrarily assign position (9) in my catalogue of properties of this model. The logically general theories required in the sentential theory are, just because of the requirement of generality, embarrassed by a lack of riches. One must suppose that anyone who asserts a sentence of the form "For all x . . ." means what he is saying, that is, that his sentence applies to everything in the world, past, present, and future. But direct observation is not this capacious. Among the constraints is the vitally important one that neither I, nor any collection of observers, can penetrate all that is past; neither can we anticipate with completely certifiable foreknowledge what will happen in the future. This is one rather clumsy way of stating what, since the work of Hume, has come to be known as the problem of induction. To raise the discussion one level of technicality we can point out that all logically general statements have reference to all that is, was, and will be; but unfortunately the actually available set of referents of such a universal generalization is observationally finite. Adherents of the sentential theory of theories must, therefore, either offer what they take to be a solution to the problem of induction, or they must presuppose a solution implicit in their philosophy of science. This is not true of falsificationists like Popper, for whom only the dagger of the present case stabbed into the willing—one hopes already dying—body of the theory matters. For any H-D theorist who thinks that the problems of confirmation theory will one day be solved, induction continues its role as the dark, indelible spot, invulnerable to the application of the grues and bleens that would render it biodegradable. (This is a taunting, ingroup, self-indulgent metaphor. Equal time should be given to Nelson Goodman's important and brilliant *Fact, Fiction and Forecast*.)

Throughout my exposition of the nine major characteristics of the sentential theory of theories I have not hesitated to suggest criticisms, nor to point up consequences of the theory that may lead to the necessity to abandon it. In the literature in philosophy of science in the twentieth century, the volume of criticism of the model has been enormous, and the types of criticism quite various. No one of these nine points has been neglected. In closing this essay, I want only to introduce two lines of criticism: one as an example of the kind of objection that can be directed at a single vitally important ingredient of the theory; the other not so much a criticism as a sympathetic suggestion of the way in which the theory can be altered to make it acceptable. First, I will turn attention again to the especially important reliance upon deduction characteristic of the model; second, I will try to show that something important and of continuing consequence is involved in the sentential theory's insistence that science must be about a world which it does not create—a world beyond theory.

It must be granted, I think, that in stressing the deductive aspect of science, the sentential theory does indeed correctly mirror a major aspect of the process of scientific inference. As I have mentioned, compliance with the regulative ideal of seeking greater and greater generality appears characteristic of the history of all sciences, and is denied as an ideal only by those for whom—in the most narrow positivistic view—science is a mere description of what has already occurred, a science devoid of genuine explanatory value. Although the sought-after model of scientific inference is deductive, however, the model is seldom historically attained, nor is it free from certain defects.

It was Pierre Duhem who clearly and disturbingly pointed out that the history of science—even the history of physics, where one expects the greatest mathematical precision—yields very few actual cases in which a new general law can be shown logically to imply earlier, less general laws. Indeed, Duhem had the audacity to try to show that not only are some important laws involved in the Newtonian synthesis not deductively implied by Newton's system, but that certain of these laws are logically incompatible with Newton's inverse-square law! (It is not at all easy to show the required contradictions. Leaving aside always troublesome details, the scenario reads something like this: We would need to show that predicted values for various applications of Newton's law differ from those for, say, Galileo's accelaration-rate law to an extent that cannot be accommodated by the calculus of standard observational error and cannot otherwise be standardized in measurable ways by means of non-trivial changes in theory.) Whether or not Duhem is right in this challenge, it is certainly a thesis that cannot be ignored, especially given that historically few, if any, early attempts were made actually to derive Kepler's and Galileo's laws from Newton's law. Only in the subsequent history of physics was the Newtonian synthesis worked out and, in a sense, vindicated. If, in addition to considering laws as mathematical idealizations, we take into account that they are usually interpreted in the context of some general ontology for physical objects, then Duhem's point about logical incompatibility is transformed into one about the difficulty of the *logical application* of idealizations falling short of ontological realization. Thus viewed, Duhem's dagger points at laws like Galileo's and Newton's *within* the bodies of their own systems. Galileo's law requires for its application the existence of vacuua; a firm believer in the view that physical reality is—in its interpretation—particulate, Galileo had to accept a mathematical idealization that contradicted his ontological proclivities. Newton's law is left in something of the same quandary: If the law is true, as everyone believed, then there is action at a distance, a concept clearly at odds with Newton's accepted ontology for mechanics that taught that all physical action is initiated through collision or impact.

One can of course suggest that simple deduction does not depend upon such subtleties of ontological interpretation. And, of course, simply and abstractly it does not. But remember that the sentential theory has locked itself not only into the harness of deductive structure, but also into the required constraints on the *applicability* of a theory, constraints imperatively needed exactly to insure that the deductions—where they obtain—*are not factually empty*. Insistence upon the testability of theories and hence upon the requirement that theories have empirical content commits the H-D model to both forms of Duhemian worry about the deducibility requirement.

Some of us might be content to give up the deducibility requirement on sentential (and other) theories on the kinds of grounds Duhem supplies. Many philosophers, however, are not readily persuaded by arguments that appeal to historical considerations (just because it is that way doesn't mean that it *should c: must* be that way), or to historical considerations alone. For the most part, I have been dealing with hypotheses or laws as premises from which consequences logically follow (or fail to do so). But we also have seen these laws in their role of *consequences* of arguments, as in the cases where *modus tollens* validly applies. In this neighborhood lurks an evil logical genius who cannot, I think, even in the best of seasons be exorcized. That evil genius is the menace of logical deduction running in the wrong direction. We want a body of observational data—expressed sententially in the accepted observation language—that implies a certain hypothesis. But we are clear on the point that such implication cannot be deductive; it can only be confirmatory (we are not here dealing again with Popper's falsificationism, a position taken to have been shown to be inadequate above). But, as we have seen, no body of confirmatory evidence can once and for all establish the truth of its putative law. It follows that any body of observational data can be taken as supporting as many hypotheses as we wish, even in those cases where the hypotheses have different empirical content. Put less esoterically: For any body of observational data O, there is a perhaps infinite set of hypotheses H, any one of which is supported *confirmationally* (notice not *logically*) by O, even if some members of the set H are in the strict sense logically incompatible with one another. This point is not novel; Wesley C. Salmon has investigated this and other shortcomings of the H-D model in his excellent *The Foundations of Scientific Inference.* For Salmon, the problems are solved by reconstructing the inferential structure of science along Bayesian lines (look this one up for yourself!).

For us, in the present context, the problem is this: If there are residual and damaging problems in Popper's program, and confirmational logical considerations also lead us into trouble, where do we turn? Well, I think the first turn is one that might have suggested itself to us all along: What the sentential theory of theories attempts to do is just not fully possible. On the one hand, this theory strives to characterize science as (ideally—note the normative strain) preserving the central features of deductive systems—science is applied mathematics, at least in the best circumstances. On the other hand, the H-D model insists that the formalism preserving deductive logical connection also must be connected with "reality," where the meaning of this reality is given without the intrusion of theory. I have led you to believe that there are really fundamental problems about the "given," while at the same time trying to argue that some vestige of this idea must be preserved if we are to have sciences whose truth is not simply self-certifying.

The problem here—and I take this to be *the* problem of the sentential theory of theories—is that the sentential theory is the heir of historically determined philosophical *empiricism*. (Of course, this is not going to be the whole story: No contemporary philosophy is so circumscribed that the actual amount it inherits from a certain paternal group can be exactly ascertained.) Something must be said, however, about the links, if not in history, then assuredly in epistemology and ontology. Nevertheless, the damage done by earlier philosophical empiricism to the contemporary sentential theory is not far to seek. The burden of sin of the twentieth-century H-D model of science is that it forces upon us a single ontology that is essentially (in very broad terms) *phenomenalist* in nature. We have already seen part of what this means in our discussion of observation languages. What an observation language gives us is a set of terms whose *meanings* are theory-independent and based on observations. This kind of language, further, is to provide us with *evidence* for or against a certain theory being put to the test. A fascinating and unobtrusive slip occurs: For human beings observation is some kind of seeing (experiencing, in a larger sense). Which means, for openers, that scientific observing must be reducible to seeing, hearing, or otherwise *sensing*—where in the best circumstances, for the sake of the *publicity* of scientific findings, the sensing is done by an *instrument*, a contrivance full of mathematics and packed with assumptions about the nature of human experiences as correlated with certain hypothetical constructs of science. The "blue" that normal human beings see is replaced by correlation with a range of colour bands on the spectrum represented as units of "visible" light, and there are further moves that reduce these intervals on the spectrum to what are called "angstrom units," *numbers* of a very special kind. What is going on here is full of significance for the nature of observationality as we must understand it in science.

If the *meaning* of observation terms must finally refer to your sensory experiences and mine, then science is in a shambles of different vivid and dull, sharp and smooth, lively and vapid sense experiences. No generality here, no apparent lawlikeness—most certainly not, if we are constrained to consult the experiences of *everyone*. There is no everyone (note the practice in all sciences of selecting a "sample" from a "population"). The move is obvious: There is no theory-independent observation language, or at least not one whose terms are given a common meaning by the sensory experiences of human subjects. At one extreme (often noted), each experience is given a theoretical

53

colouring by the language by means of which that experience is verbally expressed. At the other extreme—already noted—the experience is correlated to numbers discoverable by all. The point is that in both cases observation is not theory-neutral, but crucially theory-dependent. We arrive at a situation that seems to break the back of the sentential theory. There is no observation language whose terms are given meaning apart from theory; hence there is no "factual" test of a theory; hence two competing theories might be what Feyerabend calls "incommensurable"—sharing no common observation terms, the theories cannot in any logically proper sense be called competitors, even though they pretend to explain the "same" domain of facts. Whewell's dream of consilience is dismissed: The epistemological desire for crucial experiments that will decide once and for all between logically disparate theories remains forever unfulfilled.

Enchanted with the neatness of the logic of crucially deciding experiments, and as yet unaware that Duhem's contextualism was eventually going to come to the forefront of my thought and in some ways dominate it, I reacted against the problem of incommensurability by attacking Feyerabend's early version of this epistemological extreme. I still believe that the motivation behind the criticism was correct—I still want to find satisfactory ways of deciding between competing theories and of accounting for scientific progress. But I am now convinced that part of what Feyerabend attempted in his early work is in principle—if not in detail—correct. I am persuaded that we all operate within the confines of what the late Everett Hall called the "ego-centric predicament": Our givens are all interpreted in some system or other, otherwise we could not understand anything, appreciate anything, love anything, worship anything. But then we have known this, as I suggested earlier in this essay, since Kant; but Kant's arguments are often so obscure that we prefer the simplicity of the sentential theory. In any case, what was right about Feyerabend's early work was his recognition that we need a new concept of evidence that does not rely on empiricist theories of perception if we are to have ways of judging the acceptibility of scientific theories.

Feyerabend's general point was that what we need in order to establish the meanings of observational (experiential) terms is not some obscure appeal to common perception, but a pragmatic criterion of the acceptability of certain terms into our language. The insight involved in this "pragmatic theory of observation" is that the kind of evidence we need to test theories (and now we are free to regard theories as being not just sets of sentences, but also policies, ways of acting, etc.) is pragmatic, involving decisions resulting in actions of certain kinds. As a theory of meaning, Feyerabend's account is very defective; but as a suggestive hypothesis about the nature of evidence, it is very rewarding indeed. The sentential theory has done a real service in endeavoring to rescue scientific realism. Feyerabend has shown that what it attempted to rescue was the *wrong kind* of realism. The arguments point conclusively to the conclusion that scientific experience based on a sensory given is mistaken. The "tests" of scientific theories cannot any longer be construed along earlier realist lines. This means, I think, that a radical new thesis must be introduced: What we had earlier regarded as *extra-evidential* measures of the epistemic worth of theories must now come to be recognized as *evidential* measures in the context of a vastly revised notion of what constitutes "evidence."

On this view the simplicity of a theory (not its aesthetic properties, but the number of claims that it can make from few assumptions) will be counted as part of evidence for or against the theory. Likewise, the success of a theory, with regard both to its correct predictions and to how well it *works* when put into practice, will count as evidence. This list of pragmatic-evidential conditions can be extended. The general point is that we come to see that science is a goal-directed activity and, like any such, seeks to achieve the realization of certain utilities. Seen in this light, the question of whether or not our sciences are acceptable when compared with other forms of activity seeking to realize similar objectives is answerable by consideration of the results obtained by application of the theory. Another feature of this thesis forces us to view science as a more sophisticated extension of other, more primitive methods of problem-solving. What the earlier American pragmatists would have called the "cash value" of a science is at stake, and, of course, that depends finally upon how much risk capital one is prepared to venture. The value considerations will, of course, be quite sweeping; no longer will science and value (including ethical value) be seen as distinct—it matters to people whether or not they live in the vicinity of a cyclotron.

What I am sketching as a way out of the difficulties latent in the sentential theory may also give us a pragmatic way of specifying a kind of common context (not language) of observation. A Hopi Indian and I may not speak the same language (and hence may not share the same "theoretical" ontology about the world), but we are, I think, inclined to react in the same way in the presence of an oncoming locomotive, even if the Hopi takes it to be a newly discovered kind of animal and I take it to be a complex piece of machinery. There is this given-in-action that has Lewis' given as its proper ancestor. It may even be that a psycho-physiological account can be given of such reaction patterns. If so, then the "commonness" of experience will again be achieved by means of a correlation between physiological processes and purposive forms of human action. Feyerabend recognized something like this pragmatic dimension of scientific activity.

I have no space left in which to develop my sketch of the new (old) pragmatic realism. A final word must be said, however, on behalf of the sentential theory of theories. It is an important characteristic of philosophical theories of great scope that they are problematic—some might say ambiguous—at just those crucial points where one might hope for the greatest degree of clarity and precision. I take this to be a virtuous shortcoming, or an enlightened vice, of all such theories. The gaps in the arguments of Plato in part made possible the work of Aristotle. The unguarded tenacity of Kepler and Einstein highlighted the problems urgently needing to be solved. It is to the credit of those who worked so hard to preserve various aspects of the sentential theory that they left so much that is unclear and historically inapplicable: They created paths to follow, not sacrosanct dogmas to kneel down to. That is the significance of all philosophy (including philosophy of science): Honestly contemplated, no one knows where a given path will lead. What is important is to choose a path and learn what one can'learn by walking it.

Bibliography

Butts, Robert E., "Feyerabend and the Pragmatic Theory of Observation," *Philosophy of Science*, XXXIII, 4 (1966).

———, *William Whewell's Theory of Scientific Method*, ed., (Pittsburgh 1968).

Carnap, Rudolf "Testability and Meaning," *Philosophy of Science,* 3 (1936) 4 (1937).

. ———, "*Foundations of Logic and Mathematics*," 1, 3, *International Encyclopedia of Unified Science* (Chicago 1939).

Duhem, Pierre, *The Aim and Structure of Physical Theory*, trans. P.P. Wiener (New York 1962).

Feyerabend, Paul, "Problems of Empiricism I," *Beyond the Edge of Certainty*, ed., R. Colodny (Englewood Cliffs, N.J. 1965).

Goodman, Nelson, *Fact, Fiction, and Forecast*, 2nd ed. (Indianapolis 1965).

Grunbaum, Adolf, *Philosophical Problems of Space and Time*, 2nd ed. (Dordrecht, Holland 1973).

Hempel, Carl, *Philosophy of Natural Science*, (Englewood Cliffs, N.J. 1966).

————, *Aspects of Scientific Explanation and Other Essays in the Philosophy of Science*, (New York & London 1965).

Lewis, C.I., *Mind and the World Order*, (New York 1956).

Nagel, Ernest, *The Structure of Science*, (New York & Burlingame 1961).

Popper, Karl, *The Logic of Scientific Discovery*, (New York 1959).

Quine, W.V., *From a Logical Point of View*, (Cambridge, Mass. 1953).

Salmon, Wesley, "The Foundations of Scientific Inference," *Mind and Cosmos*, ed., R. Colodny (Pittsburgh 1966).

Traditional Philosophy of Science: A Defense

JAMES W. VAN EVRA

Philosophy of science is a discipline primarily concerned with the critical examination of the myriad assumptions upon which science itself rests. Within its limits, many attempts have been made to provide a reasonable account of the epistemological and ontological underpinnings which render science possible. In fact, since the closing decades of the nineteenth century, the pace of these attempts has propelled it into becoming one of the most active fields within the total philosophical enterprise. During the intervening period, we have gained, if not an unshakable knowledge of the foundations of science, at least a vastly improved understanding of the many possibilities toward that end, together with an understanding of the conceptual pitfalls which many of those possibilities conceal.

The intense interest in probing the groundwork which underlies science is the result of two broad patterns of influence: First, by the mid-nineteenth century, science was showing the first distinct signs of coming apart at the seams. Where science had been relatively quiescent, basking in the glow of Newtonian rectitude, by about 1880 ominous things were beginning to happen: While physics was experiencing the onset of what were to be chronic difficulties in maintaining the Newtonian landscape, mathematics was having the first taste of paradox, and thereby a glimpse of the limitations inherent in mathematical reasoning, limitations neither previously seen nor suspected. Biology was being reshaped in Darwin's wake, and chemistry was going through the extended pangs of its rebirth earlier in the century. All, in other words, was far from quiet.

When science is in such a state of turmoil, there is often a marked increase in concern with foundational problems. When science goes berserk, that is, many of its practitioners begin to wonder about the cogency of the foundational assumptions on which their enterprise rests. Much like the military maneuver of retreating to safer ground, science attempts to re-establish some firm conceptual underpinnings from which to assess the events which wrack it. Since this return to foundational questions lies at the heart of philosophy of science, the sudden upsurge in interest in arriving at some relatively unshakable perspective on scientific methodology just when science was showing stress is more than mere coincidence.

The other major influence on the course of twentieth-century philosophy of science stems from problems within philosophy itself.

Since the mid-eighteenth century, various versions of a theory generically referred to as "idealism" had enjoyed a significant following among thinkers. This theory, again enjoying prominence in Europe in the late-nineteenth century, asserts the ultimate dependence of the real and knowable upon mind. Critics regarded it as a direct threat to the idea that cognitive knowledge is basically objective, i.e., that the legitimacy of knowledge claims is independent of the way in which the mind views the world. Objectivity in this sense implies the availability of means for testing knowledge claims which are independent of an exclusive reliance on the operations of minds.

Not surprisingly, those interested in finding a firm footing for science were among those most opposed to idealism in any form. The mere suggestion that the way science works, i.e., its methodology, might be primarily dependent on the mind, was met with scorn and derision. Instead, the ideal envisaged by the majority of turn-of-the-century philosophers of science was the articulation of a methodology for science which would indisputably show that it rests on a foundation impervious to the foibles of human mentation. Otherwise, it was felt, science would be foundationless—i.e., changeable on the basis of fad and opinion. Under such circumstances, real, rather than merely apparent, scientific progress would be impossible.

This predominate group of philosophers of science, called "logical positivists," or "logical empiricists," offered, as an alternative to what they saw as the meaninglessness of the metaphysical views spawned by idealism, a particularly severe version of another philosophical theory, empiricism, the generic view whose central contention is that knowledge is ultimately experience-dependent. In the positivist's version of empiricism, the legitimacy of knowledge claims could be tested by tracing them to their basic experiential antecedents, which were not themselves dependent on our modes of thought. Basic experience so construed provides absolute access to the world and hence a way of grounding the manner in which scientific hypotheses can be validated.

II

The foregoing, no-frills account of the setting in which twentieth-century philosophy of science developed is designed to provide some background for consideration of the main issue here. Logical positivism and its descendants (to which I collectively refer as "traditional" philosophy of science) held nearly a monopoly in the field from its inception early in the century until well after the Second World War. During this period, there was a simple presumption, widely held, that although traditional philosophy of science faced some nagging prob-

59

lems, in the long run they would disappear and positivism, in some form, would prevail.

A decade or so ago, however, some serious opposition to the traditional view began to appear from a group (which includes, most prominently, Thomas Kuhn, Paul Feyerabend, and N.R. Hanson) usually identified as subjectivists in the philosophy of science. The point of their criticism is that traditional philosophy of science is wrong-headed and, in fact, serves more as a detriment to science than an aid. They raise the spectre that philosophy of science in its zeal to provide foundations for science has become hermetically sealed off from science itself. Rather than aiding the advance of science, they depict philosophy of science as isolating itself by concentrating on technical problems which are generated from within the discipline.

In what follows I offer reasons for thinking this view is simply wrong. My unhappiness with the subjectivist attack on traditional philosophy of science stems from what I take to be a major inadequacy in the subjectivist conception of science itself. For the subjectivist, science is simply equated with the activities of its practitioners. Subjectivism, which developed in part as a result of a renewed interest in the history of science, portrays science as a highly visible, strongly social, thoroughly human enterprise which revolves primarily around the activities of scientists. For the traditional philosopher of science, on the other hand, *science itself* is held as an *object of scientific scrutiny*. The product of traditional philosophy of science often bears little resemblance to the subjectivist's conception of science, in just the sense in which the scientist's description of something may bear little resemblance to its outward appearance. To the subjectivist, traditional philosophy of science is confining and irrelevant, since it does not aid and abet science *as he sees it*. What the subjectivist offers, on the other hand, seems *prima facie* to be a "liberated" view of science, since it avoids the formal structures, paradigms, and technical distinctions which abound in the traditional conception of scientific methodology. While shunning such devices, the subjectivist offers instead a picture of science "as it really is," full of audacious goings on, which at their best seem totally resistant to uniform characterization.

In fact, precisely the opposite is the case. The subjectivist view, while apparently liberating, is really confining: It limits the perspectives available in which to view science. And, since it resists attempts to characterize science, it replaces solidly based knowledge of the nature of science which appeals to intuition as to what is good science, and what is not. The traditional approach, while *apparently* confining, is, in fact, the truly liberating one. Out of the line of development from the early positivism of the group known as the Vienna Circle, that is,

60

has emerged, through a trail of gaffes, errors and misdirections, a new theory of science, in terms of which science is coextensive with the total field of cognitive knowledge. Science is now seen as resting on no absolute foundation of unassailable knowledge, i.e., as being composed of no statements which are immune to revision. The same general theory also provides a comprehensive account of ontological commitment which not only involves considerations of how existence claims in science are handled, but places science in the total ontological picture as well. These points, to be explained more fully later, lead to an inescapable conclusion: Traditional philosophy of science has been a liberating influence by providing a coherent, comprehensive, rational account of the foundations of science.

Besides the generally confining character of subjectivism, some of its more specific features are troublesome. First, in their zeal to criticize traditional philosophy of science, subjectivists concentrate on a brand of philosophy of science (pure positivism) which had vanished and been replaced by its more liberal offspring before subjectivism ever appeared. Also, if philosophy of science stands in relation to science as a science stands to its subject matter, then the demand made by subjectivism that philosophy of science participate directly in the activity of science is akin to what Richard Rudner calls the "reproductive fallacy." The fallacy involves the mistaken belief that the business of science is not to characterize its subject matter, but to capture it *in toto*, or reproduce it within the observer.

In discussing the central issue raised in the dispute between these theories, I shall concentrate primarily on the work of one of the most persistent and articulate (and impassioned) critics of the development of traditional philosophy of science, Paul Feyerabend. As a subjectivist, Feyerabend's criticisms are particularly important in light of his profound knowledge of the history of science. In what follows, I begin with a brief account of the main aspects of Feyerabend's position, followed by a criticism of it along the lines already laid out.

Feyerbend's conception of science involves detailed historical analyses of various episodes which, he believes, illustrate important characteristics of science. His basic and oft repeated conviction is that, instead of pursuing the often highly technical and formal concepts common in traditional philosophy of science, one should concentrate on contemporary and historical, primary scientific sources. But though he shies from offering an all-encompassing characterization of science, some features which he considers particularly significant emerge in the course of his work. They include the following:

First, science, *really* to work should avoid the isolating influences which constrain it and render it dogmatic, authoritarian, and narrow-minded. (The sort of science characterized by Kuhn [see bibliography] as "normal" serves as an example of the constraint he has in mind.) Once barriers are raised to the pressure of external influences, science automatically loses its resilience and vitality; it becomes inward-looking, concentrating on problems exclusive to the discipline. Its chances for survival as a social institution are thereby diminished: It becomes a less interesting and hence less viable activity. Feyerabend's ideal, on the contrary, is a situation in which science is not a preserve open only to the "properly initiated," but is carried on along a broad front of common activity which includes virtually all areas with a potentially direct bearing on science itself. Breaking down the barriers eliminates the conformism which is commonly generated by them and produces a much more vital, critical, human intellectual activity. As an example, Feyerabend suggests Galilean science: As he describes it, theorizing undertaken then was subject to a vast array of not uniformly rational influences from amazingly diverse sources. The result was a cacophony of activity in which these diverse influences (including philosophy) actively *participated* in science and actually contributed to changes in it.

Secondly, again by looking at science "as it is" (or rather was), Feyerabend concludes that the quest for a methodology which incorporates rules, structures, distinctions, etc. is pointless. In fact, as he sees it, science has a special knack for breaking the most carefully laid down, the most "obvious" rules of procedure, and for functioning in a manner directly contrary to such rules. His alternative is to recognize science as a quest in which "anything goes": We are asked to recognize the full range of forces which shape the course of science, including, from time to time, outright propaganda, counterinduction, total disregard for the "facts," ad hoc hypotheses, and other activities which might seem outrageous to the methodologist bent on discovering stabilities which underly the course of science. Feyerabend's message to such a methodologist is:

> Do not work with stable concepts. Do not eliminate counterinduction. Do not be seduced into thinking that you have at last found the correct description of "the facts" when all that has happened is that some new categories have been adapted to some older forms of thought, which are so familiar that we take their outlines to be the outlines of the world itself.

A broad proliferation of theories is needed, whether ad hoc or counterinductive, as well as those which seemingly "fit the facts." As for method, well:

> The idea of a fixed method, or of a fixed (theory of) rationality arises from too naive a view of man and of his social surroundings. To those who look

at the rich material provided by history, and who are not intent on impoverishing it in order to please their lower instincts, their craving for intellectual security as it is provided, for example, by clarity and precision, to such people it will seem that there is only one principle that can be defended under all circumstances, and in all *stages* of human development. It is the principle *anything goes.*

Yet another component in Feyerabend's position is his more general epistemological convictions. These include a strong naturalism—"Knowledge is a part of nature and is subjected to its general laws"—together with a strong Hegelian bent to that naturalism:

> The laws of dialectics apply to the motion of objects and concepts as well as to the motion of higher units comprising objects and concepts. According to these general laws, every object participates in every other object and tries to change into its negation.

His epistemology is as foundationless as his methodology. Any attempt to lay down a rock-bottom foundation on which knowledge might rest is doomed—there simply is no ascertainable starting point for knowledge—and to presume the existence of one is to harm future inquiry needlessly by tying it to what merely appears to be a point of stability. In particular, Feyerabend believes that all of the attempts to press for an epistemology based on a distinction between raw experience on the one hand, and conceptualization on the other, (a distinction widely held by positivists) miss the point. There is simply no unsullied (i.e., raw) experience. What we call experience is for Feyerabend dependent on all sorts of theoretical pressures which cannot be stripped from it.

From the foregoing, it is not particularly difficult to predict what Feyerabend thinks about philosophy of science in its traditional guise: If it does not visibly participate in and *change* science, reject it. Hence, to analyze such abstract devices as paradigms of theory structure and idealized accounts of scientific explanation (which is precisely what traditional philosophy of science does) is only of value if such devices can be seen to be doing something in the course of scientific activity.

On this basis, twentieth-century (and in fact, all post-Newtonian philosophy of science seems to be a dismal failure which could be scrapped with no net loss to science itself. Feyerabend sees philosophy of science as having severed any relation it might have had with science and characterizes it as a perfect example of a self-centered, strongly conformist activity generating technical problems of its own rather than concentrating on science itself. Thus it has lost, he says, "any chance of adding to our knowledge of the world." Its "formal castles" have "nothing whatever to do with what goes on in the sciences. There is not a single discovery in this field (assuming there have been discoveries) that would enable us to attack important scientific problems in a new way. . . ."

As if this were not enough, methodology in its present guise, Feyerabend contends, has had detrimental effects on the development of the social sciences. Taking as gospel the prescribed methodology, and attempting to construct a science on that basis has left, in his estimation, the social sciences in "sorry shape."

Specific targets of Feyerabend's unhappiness with contemporary philosophy of science are not hard to find. Philosophy of science traditionally has been committed to the possibility of distinguishing between the manner in which we arrive at hypotheses (the context of discovery), and how they are eventually tested (context of justification). Like Kuhn, Feyerabend holds this distinction to be untenable: ". . . The distinction refers to a situation that does not arise in practice at all. And, if it does arise, it reflects a temporary stasis in the process of research. Therefore, it should be eliminated." Another old distinction, that between observation terms and theoretical terms, i.e., between those terms which directly relate to observation and those dependent on theoretical context, meets the same fate: Abandon it because it does "not now play a role in the business of science." Moreover, the contention that experience is the foundation of our knowledge is an hypothesis that cannot be effectively examined if we use a method which justifies or criticises it "on the basis of experience."

Given his basic epistemological proclivities, it is also not surprising to find Feyerabend antagonistic to the idea (again, popular with the early positivists) of a pure observation language (i.e., one which directly confronts immediate experience). He refers to such a language as a myth, given the basic impurity of observation. Similarly, the idea that theories are tested by determining how well they "fit the facts" will not do either: It is possible to cite instances in science in which the facts have been tailored to fit theory. The traditional theory of confirmation also comes under attack for suggesting that theories can be judged in terms of their ability to "fit the facts." In this regard, Feyerabend is particularly scornful of the so-called "grue-bleen problem" as epitomizing what he sees wrong with traditional philosophy of science.

Finally, the use of formal logic to explicate the foundations of science falls squarely within Feyerabend's sights:

All I wish to assert is that there exists an enterprise which is taken seriously by everyone in the business where simplicity, confirmation, empirical content are discussed by considering statements of the form (x) $(Ax \rightarrow Bx)$ and their relation to statements of the form Aa, Ab, Ac & Ba, and so on and *this* enterprise, I assert, has nothing whatever to do with what goes on in the sciences.

In sum, Feyerabend is not overwhelmed by what he sees in traditional philosophy of science.

There is, without doubt, much that is immediately appealing in this view. For instance, in his historical analyses, he succeeds in capturing many central features of science as a social institution. From an historical perspective, his work is an engaging, often humorous, and always provocative account of the workings of science. Moreover, I suspect that his exhortations to pull out all the stops, to revere nothing, may serve as an inspiration to the working scientist.

But what of the remainder of Feyerabend's message? There is *apparent* appeal in the distinction which he draws between science proper and traditional philosophy of science. Here is science going pleasantly berserk, responding to seemingly wacky influences, riding roughshod over proferred rules, and refusing to suffer the boredom which characterizes illusory stability. On the other hand, we have philosophy of science: It is heavy, slow moving, insufferably technical, and, just as Feyerabend claims, bears little resemblance to science. Its problems are not science's problems. Philosophy of science *adds* problems to the conceptual scene rather than solves them, and it rests on concepts which seem detrimental to science, were they seriously to be adopted by it.

But while Feyerabend may be dead right in his description of science-as-a-human-activity, he is dead wrong in his estimation of contemporary philosophy of science. And he is wrong, as I suggested above, because of his very provincial view of the nature of science which prevents him from realizing that there *is* no essential boundary between science and philosophy of science: Philosophy of science *is* the theory of science, just as celestial mechanics is the theory of the motion of heavenly bodies. Once this kinship is admitted, as I think it must be, then Feyerabend's demands on, as well as his criticisms of, philosophy of science can be seen to be totally unsupportable.

His provincialism is apparent from a number of different considerations. For instance, had Feyerabend turned his attention to primary sources in the history of philosophy of science, as he admonishes us to do in the case of science, he would have discovered an activity of exactly the same variety as he praises in science itself: As a social activity, the development of philosophy of science from the early days of the Vienna Circle to the present is replete with the same twists and turns, the same counterinductive hypotheses, as Feyerabend finds in science itself. The same principle of proliferation has been at work and, if Feyerabend appreciates zany ad hoc hypotheses, he should *love* the rich interplay which surrounds the grue-bleen problem, rather than hold the keepers of the problem in contempt. But of course, science and the philosophy of science are not the same *sort* of thing for Feyerabend: Science is the *activity* of its practitioners, but philosophy of science is viewed entirely in terms of its *products* and how they relate to science. Had he viewed science *itself* in this manner, much of the madcap zaniness which he finds so appealing would vanish without a trace. Science so regarded would be reduced to dull statements, equations, distinctions, constants, and formal structures of the same sort which Feyerabend deplores in the philosophy of science.

That the brand of philosophy of science at which Feyerabend cavils has been a science of science from its very inception is readily apparent from even the most cursory glance at its development. One need only, as a start, look at what the members of the Vienna Circle claimed as the goal of their work. In an early article, for example, Rudolf Carnap says,

> Philosophers have ever declared that their problems lie at a different level from the problems of the physical sciences. Perhaps one may agree with this assertin; the question is, however, where one should seek this level.

Carnap's answer is,

> In order to discover the correct standpoint of the philosopher, which differs from that of the empirical investigator, we must not penetrate *behind* the objects of empirical science into presumably some kind of transcendent level; on the contrary we must take a *step back* and *take science itself as the object. Philosophy is the theory of science.*

And while the substantive concerns which surround attempts at attaining such a theory have changed dramatically since Carnap's day, the fundamental orientation of philosophy of science as a science of science has not significantly changed at all.

To be sure, the road to a tenable theory of science has been littered with the discarded wreckage of many failed attempts. These mistakes, in turn, stem from relatively few sources, all of which revolve around the assumption that a foundation for science could be found in some preferred mode of knowledge, which in turn would serve as a fixed base for the rest of knowledge and belief. It was simply assumed by many early philosophers of science that if some Archidemean point were found which could be used as the ultimate court in which knowledge claims could be tried, everything else would fall nicely into place. The choice for that epistemic base is well-known: A strongly reductive phenomenalism based on Hume's empiricism, with modifications by, among others, Mach. For several decades after the birth of traditional philosophy of science, inquiry was dominated by attempts

66

to make this empiricism work by continually shoring it up in the face of a seemingly endless series of difficulties.

The other major influence on philosophy of science throughout this period was its practitioners' very optimistic view of the logic of *Principia Mathematica*. This powerful logic looked like a truly formidable weapon to be used with effect in the attainment of a theory of science. The idea was that, on the one hand, logic could be used in the characterization of the *structure* (i.e., formal properties) of science, while reductive empiricism could serve as its substantive base. (As an instance of the level of optimisim with which logic was greeted, it is interesting to note that in the article quoted above Carnap goes on to identify the theory of science with the logic of science.)

The result of these influences looked truly formidable. The combination of extreme empiricism and mathematical logic *appeared* as if it might be just the sort of rock on which all that was worthwhile could be built—a structure with truly Cartesian properties. What actually happened fell considerably short of this ideal. From the start, attempts at establishing a principle of empirical verification, i.e., a rule relating truly meaningful expressions to conditions of direct empirical test, ran into considerable difficulty. Just as soon as one such principle was proposed, a loophole would be found in it through which legions of "meaningless" statements could march undisturbed. Yet for nearly twenty years, the quest for a single point of absolute connection between science and raw experience continued unabated. But by 1950, when Hempel wrote "Problems and changes in the empiricist criterion of meaning" which chronicled the series of these attempts, it was finally beginning to appear that the quest was in vain. Given the realization that no verification principle was forthcoming, main line empiricism still seemed to be in relatively good shape. Even if a hard and fast principle of verification was not forthcoming, it presumably still made good sense to defend empiricism by other means. If statement-by-statement verification by experience was not possible, perhaps empiricism could be sustained by reliance on an entire language which confronts experience en masse rather than in molecular bits. Translatability into such an empiricist language would then be the ultimate test of meaningfulness. Moreover, if the distinction between such an empiricist language on the one hand, and other varieties of language (including theoretical) on the other, turned out to be tenable, then all would be well. With a tightly circumscribed observation language to rely on, the basic components of science could still be rationally accounted for. Theories could be viewed as interpreted logical calculi possessing determinate formal structures, and the nagging problem of specifying the nature of the connection between the language of theory and of observation could be resolved by specifying connections which

67

were weak enough to permit theoretical language to be open-ended in the scope of its application on the one hand, and yet were restrictive enough to retain an over-all reliance on the level of pure observation on the other. But the promise of this programme did not last. By 1958, the battle was substantially lost: That was the year that Hempel wrote "Theoretician's dilemma," a remarkable article which begins with the presumption that a strict distinction between observation and theory is possible, and ends only after he has admitted that, in fact, such a distinction may not be tenable, and is, in any event, unnecessary.

Thus, by the late 1950s, the classical programme of the logical positivists had, if not completely vanished, at least shown signs of impending mortality. With the demise of the sharp distinction between observation and theory, there no longer seemed to be much point in pursuing many of the traditional goals of the movement: They rested on a putatively firm base for knowledge which now no longer seemed to exist. The positivist remnants then consisted mainly in a paradigm for theory structure with no absolute connection to extratheoretical states of affairs, a paradigm for the structure of scientific explanation which incorporated problematic components of its own (law-like-ness, for instance), a trouble-ridden theory of confirmation, and very little else.

This history of accumulating woe leading to apparent failure might prompt one to suspect that Feyerabend's attitude to the whole enterprise is essentially correct, but there is more to the story. Before proceeding to it, however, there are some significant points to be noted. First, nothing in the entire positivist enterprise was ever *intended* to be *prescriptive* for science. Consistent with the attitude that a theory of science was being developed, the positivists showed no inclination to involve themselves directly in the subject of their studies. Rather, they attempted to provide a "rational reconstruction" of science which would capture the salient features of its underlying structure while avoiding those characteristics judged to be peripheral.

Secondly, the "failure" of classical positivism was really nothing of the sort. Starting from a few tersely stated, rigidly confining assumptions about the foundations of science, the process has been one of a continual exfoliation and criticism of those assumptions, criticism of just the sort Feyerabend admires when it occurs in other areas of science. In the wake of this critical process, much has been gained. Both reductive phenomenalism and nominalism have been tried and found wanting, as has the presumption that there exist unshakable epistemic roots. As the critical process continued, the quest had an increasing—and apparently permanent—impact on broader epistemological and ontological issues.

While many of the main-line projects were playing themselves out, a number of interesting alternatives were appearing on the scene, alternatives which were in the same tradition, but which offered solutions to the outstanding problems in distinctly untraditional ways. For example—one that Feyerabend curiously ignores, but that many find particularly interesting—is the philosophy of science of W.V.O. Quine. Quine's philosophical roots are firmly planted in positivism (he did his early work as a student of Carnap), and, I suspect that as a result, at one time or another, he held some fairly extreme views (such as reductive phenomenalism, nominalism, and instrumentalism with respect to theoretical entities). But when his more mature work began to appear, many of his earlier views had been replaced by new and distinctively different ones. They include: (1) the concept that there is no essential difference between science and philosophy of science. Where this had been a sketchy presumption on the part of the early positivists, Quine lays a strong theoretical base for it by proposing that the totality of cognitive knowledge and belief, from the most mundane to the most abstractly theoretical, ". . . is a man-made fabric which impinges on experience only along the edges." This field is composed of a network of statements, more closely tied to experience at the periphery, more remote from it toward the center, such that no statement within it commands a preferred position, i.e., no statement is immune to revision. As science is identified with the totality of this field, and as philosophy of science is a proper part of the field, it becomes just another aspect of the total scientific enterprise. (2) A conception of theory in which theories are no longer the exclusive property of institutionalized science, but are now conceived as underlying virtually all of our knowledge. Moreover, theories for Quine (as for Duhem a century earlier) are sufficiently underdetermined by experience that it is possible to hold a theory come what may, since they cannot be definitively falsified. (3) A conception of experience completely disengaged from earlier forms of reductive empiricism: For Quine, there is no "fancifully fanciless medium of unvarnished news," no conceptual subbasement on which experience rests. Instead, experience starts at the level of medium sized objects, and the concept of "raw sense data" only attains significance when embedded in a theory of neural stimulation. (4) A strongly realistic ontology tied to his conception of theory: To be is to be the value of a variable in a true theory.

Quine's view is about as "foundationless" in the classical sense of absolute foundation as a view can be. Since no knowledge is accorded a preferred status, no statement is held immune from revision, and no strictures are placed on the manner in which hypotheses are generated, it seems that Quine's philosophy should be nothing short of a Valhalla for Feyerabend—just the sort of place he could hang his philosophical hat without yielding any of his conceptions about how science should be done. After all, the two seemingly agree on the proper relation between science and philosophy of science (there should be no artificial boundaries between the two). And, Feyerabend's "methodological pluralism" is simply a plea against those epistemologies which depend on preferred modes of knowing. Counterinduction, if not actively encouraged, is at least permitted in Quine's view, via the Quine-Duhem thesis. As if this is not enough, Quine was advocating in 1960 precisely what Feyerabend has been arguing for since: no unconditioned experience.

In sum, Quine's position is the very liberated offspring of very straight-laced forebears. It has become progressively clearer over the course of several decades that providing a unique foundation for science is not possible and that instead, to understand the basic underpinnings of science requires a broader understanding of fundamental epistemological issues. As steps leading to this realization, the developments of the positivist period have a considerable value of their own.

I suspect that what keeps Feyerabend from seeing anything in contemporary philosophy of science except a brand of empiricism which had largely passed away even before he had begun to publish his own work is his extremely provincial view of science. Feyerbend's entire conception of science is restricted to science as process, i.e., science as the activity of scientists. Given this picture of science, there is little wonder that he might miss Quine entirely, regardless of the significant field of compatibility between them. Though Feyerabend has been philosophically scooped by Quine, the difference in orientation to science will not permit him to see it.

But more importantly, had Feyerabend paid sufficient attention to developments in contemporary philosophy of science, he might have realized that aspects of Quine's theory prove troublesome for his own view. If science and philosophy of science differ only in subject matter, then what we expect from them should not differ in kind either: A theory of science should relate to science in just the same way that a science relates to its subject matter. That relationship, moreover, does not require that a theory, to be seriously considered, must be true to the *appearance* of things which the theory is about. It is perfectly appropriate that Celestial Mechanics, for instance, ignores many of the visible characteristics of, say, Jupiter, and instead considers Jupiter as a point-mass for purposes of the theory. Regardless of how one views it, it is plain that science has never felt obliged to convey a sense of the "full reality" of the thing studied; rather, it has always concentrated on selected characteristics while ignoring many others. Were it any other way, science would simply vanish. But if botany can miss the leafy splendor of the shade tree and pass directly to a study of, say, the cellular composition of its bark and so produce a theory which entirely avoids the appearance of the tree as an inhabitant of the world of medium sized objects, then there should be no cavil at philosophy of science's missing the hustle and bustle of science in the hands of scientists when *it* concentrates on characteristics of science not immediately apparent to the eye. Insofar as it shares the same cognitive field, philosophy of science is no more constrained to concentrate on the visible aspects of science than science is constrained to concentrate on the visible aspects of reality.

This is the heart of the subjectivist failure—the inability to see the point of much of what has gone on in contemporary philosophy of science because their conception of science is restricted to scientific activity in the traditionally conceived realms of institutional science (i.e., physics, chemistry, etc.). As an instance, consider Feyerabend's contention that the justification-discovery distinction should be ignored, since it doesn't arise in practice (or, more appropriately, should not so arise). This criticism illicitly assumes that philosophy of science is locked into the purely visable "practice" aspect of science, and if what philosophy proposes does not show up in that arena, then it is worthless. But if philosophy of science is on a par with the rest of science, then this is like demanding that science itself should ignore all states of affairs which do not show up in the "practice" of reality, i.e., are not apparent in the goings on in the world of medium sized objects. If followed, that restriction would effectively eviscerate science. It would lose, in one fell swoop, many limiting and abstract states of affairs, constants with values which have not been completely computed, etc. Under this constraint, that is, science would disappear.

Viewed as theoretical devices, the *only* appropriate question concerning distinctions like discovery-justification, as well as the rest of the devices found in contemporary philosophy of science, is whether they are supportable, or not. That question, in turn, is precisely what philosophy of science has been concerned with for more than half a century. In this light, the contention that the activities of philosophy of science are not appropriate to science has the look of pure prejudice; it is one area of knowledge looking at another and saying, "Since it does not do me any apparent good, it should be eradicated."

The subjectivists' mistake is closely akin to what Richard Rudner calls the "reproductive fallacy" which involves the belief that an understanding of aspects of reality necessarily involves active partici-

71

pation in, or reproduction of, the full richness of that reality. The subjectivists' demand that, in order to deal with science, philosophy must participate in it "as it really is" (i.e., as it appears, currently or historically) is reminiscent of the claim that, to really understand tornadoes, one must stand in the path of one, or, truer to life, that to understand martyrdom, one must empathically model the mental states of the about-to-be martyr. For the philosopher, the subjectivist's one-dimensional model of science similiarly demands that he become a direct participant in the ongoing stream of scientific activity—there being literally no other alternative. And when the philosopher does *not* participate in the visible activity of science and instead erects odd-looking formal structures, there is little wonder that the subjectivist, bent on reproduction, would reject such activity as unproductive.

I suspect that the reproductive fallacy also lurks around the debate concerning the incommensurability of theories. The subjectivists, by relying exclusively on science as process, have adopted the position that incommensurability between theories makes real communication between them impossible. Hence, changing from one theory to another requires something more akin to religious conversion than to a rationally-based process of change. Once again, the one-dimensional model of science will not admit enough depth in science for a basis for theory comparison in the face of incommensurability.

III

Thus, subjectivism will not suffice as a theory of science. While its concentration on those features of science which are closest to hand robs it of a large measure of initial credibility, subjectivism is ultimately an inadequate basis for explaining the foundations of science: By forcing science into a one-dimensional mould, it adopts a relatively cumbersome perspective and must construe science differently from other things in the universe. Moreover, the subjectivist assumption that it is permissible to plumb the theoretical depths of everything except, science itself is an article of faith never exposed to rational scrutiny.

On the other hand, from the Vienna Circle to the present, traditional philosophy of science possesses the scope and simplicity which subjectivism lacks. It has been an ongoing series of attempts to attain a sound theory of science in a manner which does not differ in kind from attempts made to explain anything else. And, if traditional philosophy of science portrays science in a way that diverges from even the scientist's own immediate conception of his enterprise, no harm is done. The ultimate test of the adequacy of traditional philosophy of science, like that of any other theory, is how well it serves as a basis for understanding the functioning of science. To accomplish this, it simply is not necessary that we treat science exclusively in its most familiar aspect. We are no more compelled to maintain a familiar picture of science when we theorize about it, than we are compelled to concentrate on the taste of the apple while subjecting it to spectrographic analysis.

References

Ayer, A.J., *Logical Positivism*. Glencoe, Illinois: The Free Press, 1959.

Carnap, Rudolf. "On the character of philosophical problems," in *Philosophy of Science*, vol. 1, no. 1. Baltimore: Williams & Wilkins, 1934.

Edwards, Paul. The *Encyclopedia of Philosophy*. New York: The Macmillan Co. and the Free Press, and London: Collier-Macmillan Ltd., 1967.

Feyerabend, Paul. "Against method," in *Analyses of theories and methods of physics and psychology* (Minnesota Studies in the Philosophy of Science, vol IV). Minneapolis: University of Minnesota Press, 1970.

Feyerabend, Paul. "Philosophy of science: a subject with a great past," in *Historical and Philosophical Perspectives of Science* (Minnesota Studies in the Philosophy of Science, vol. V). Minneapolis: University of Minnesota Press, 1970.

Goodman, Nelson. *Fact, Fiction & Forecast*. Cambridge, Mass.: Harvard University Press, 1955.

Hempel, Carl. *Aspects of scientific explanation and other essays in the philosophy of science*. New York: The Free Press, and London: Collier-Macmillan Ltd., 1965.

Kuhn, Thomas S. *The structure of scientific revolutions*. Chicago: University of Chicago Press, 1962.

Quine, W.V.O. "Two dogmas of empiricism," in *From a Logical Point of View*. 2nd revised ed. New York: Harper & Row, 1961.

Quine, W.V.O. *Word & Object*. New York and London: John Wiley & Sons, and Cambridge Mass.: M.I.T. Press, 1960.

Rudner, Richard. *Philosophy of social science*. Englewood Cliffs, New Jersey: Prentice-Hall Pub. Co., 1966.

The Philosophy of Biology

MICHAEL RUSE

I shall not in this paper try to discuss every topic in the philosophy of biology. Apart from anything else, the subject is growing at such a rate that the results of that kind of approach would be dated almost as soon as they appeared. I shall rather concentrate on one topic—probably the most important—and hope thereby to give the reader an adequate flavour of what the philosophy of biology is all about.

1. Biology and Physics

To what extent can biology be said to be like the physical sciences? For one reason or another, physics is the pace-setter in the sciences: One tends to think of Newton's achievements in astronomy (or, perhaps Einstein's work), of the wave theory of light, and so on. And because of these great successes, and because also many of the men who first developed philosophy of science—men like John Herschel and William Whewell—were themselves physicists, physics has come to be regarded as the paradigm science against which all others should be judged. This then raises the question of how other sciences, biology in particular, fare when judged against the ideal of physics. Are they, indeed, essentially like physics, and if not, how do they differ? And if we agree that they differ, what then should be our conclusion—that a science like biology aspires to the same criteria as physics, but falling short, shows itself to be immature? That biology fails so abysmally that it has no right to the name of "science"? Or should we perhaps conclude that the whole enterprise is misconceived—that biology is first-rate biology and not second-rate physics; that biology has its own non-physical critera and standards?

Nearly all of these options have been taken up at one time or another by philosophers. Many of the results of their investigations are either referred to or discussed in the suggested readings at the end of the chapter. I shall here try to show how one philosopher—myself—sets about tackling the problem.

2. The Nature of Physics

To start, two preliminary matters need attention: To determine the precise criteria and standards of physics on one hand, and just what we mean by "biology" on the other. For the sake of brevity, putting on philosophical hob-nailed boots and trampling cheerfully over hundreds of subtle philosophical distinctions and arguments, I make the following suggestions about the nature of physical theories: First, they are axiomatized. One starts off with a number of initial premises or axioms, and everything else is shown to follow deductively from them—everything else is shown to be somehow "contained" in the premises. Secondly, scientific theories are composed of laws, universal statements about the world, which are believed in some way to be necessarily true. Both of these criteria—often linked by speaking of theories as "hypothetico-deductive" systems—can be illustrated by Newtonian astronomy. On the one hand, Newtonian astronomy is an axiom system. One starts with a number of premises, for example, Newton's law of gravitational attraction (namely, that bodies attract each other with a force which is indirectly proportional to the square of the distance between them), and from these one can deductively infer conclusions, for example, Kepler's first law of planetary motion (namely, that planets describe ellipses with the sun at one focus). The statements of the theory, like Kepler's law, are thought to be universally, necessarily true—all planets describe ellipses, and they *must* describe ellipses. If a new planet were discovered, as indeed happened after Newton first formulated his theory, then it would be found to describe an ellipse.

Two more points can be made about physical theories. First, those usually considered the best are found to be "consilient": That is, from the axioms' statements, many different areas of knowledge can be explained. Newton's theory, for example, explains the motion of the planets, and at the same time it can be applied to the phenomena of the tides, showing that they are a function of the varying distances (and hence gravitational attractions) of the sun and moon relative to the earth. Usually, there is taken to be a converse relationship between explanation and confirmation. Newton's axioms explain in many different areas. Conversely, that the theory can explain in these many areas is taken as evidence of the truth of the axioms. It is felt that nothing which is entirely false could have such wide explanatory power. (Obviously the coming of relativity theory modified one's views both about the explanatory power and about the confirmation of Newtonian astronomy; but for the sake of argument, we can allow that the theory explains and is confirmed within approximate limits.)

Secondly, let us note what physical theories do *not* do: explain in terms of functions, or ends, or goals. It makes perfectly good sense, for example, to ask what function the china receptacle on the floor of my office serves. It is my teapot, which I use several times each day with the aim or end of producing a nice cup of tea. On the other hand, no physicist would think much of a question about what function the moon serves. Admittedly in the past natural theologians were prepared to ask and answer such questions—the moon illuminates the earth at night-time—but a physicist would rule out such a question as improper. The objects of physics simply do not have functions. Understanding in terms of functions or ends is usually spoken of as "teleological," and it is often thought of as in some way referring to the future—I understand my teapot in terms of the future ends it will serve.

3. Evolutionary Biology

Turning now from physics, what are we to consider as biology? Once again for brevity, let me seize on one theory which is most definitely biological: the theory of organic evolution through natural selection. If this is not biological, then nothing is. We ought, however, to make a distinction. The theory first appeared in Charles Darwin's *Origin of Species* in 1859. Nothing remains the same, and in the past century the theory has undergone considerable refinement and development, particularly from the influence of Mendelian genetics. Therefore, we must distinguish between the early theory and the modern theory. Having done this, we can in earnest set to philosophizing about biology. In particular, given the four criteria I isolated from physics, to what extent are these applicable to the early and modern theories of evolution?

4. The Structure of Darwin's Arguments

Let us begin with Darwin's theory of evolution as we find it in the *Origin*. Fairly obviously, the crux of the theory is the mechanism of evolutionary change, namely natural selection. It was Darwin's belief that in each generation only a certain percentage of organisms will survive and reproduce, that for some reason which he did not understand organisms are never quite similar to each other, and that the success of the successful organisms is in part a function of the characteristics which they possess—characteristics not possessed by the unsuccessful organisms. Thus, argued Darwin, we get a "natural selection" of the successful characteristics—so-called "adaptations"— and over a number of generations this will build up to full-blooded evolutionary change.

Of particular interest to us here is that Darwin did not just drop natural selection into his theory cold. First, he argued for something which he called the "struggle for existence," incorporating the idea that not everything will survive and reproduce. He then went on to use the

76

struggle to argue for natural selection. Perhaps the quickest way to test Darwin's theory against our criteria will be to look more closely at his method of argumentation and, in particular, the argument he provides in support of the struggle for existence:

> A struggle for existence inevitably follows from the high rate at which all organic beings tend to increase. Every being, which during its natural lifetime produces several eggs or seeds, must suffer destruction during some period of its life, and during some season or occasional year, otherwise, on the principle of geometrical increase, its numbers would quickly become so inordinately great that no country could support the product. Hence, as more individuals are produced than can possibly survive, there must in every case be a struggle for existence, either one individual with another of the same species, or with the individuals of distinct species, or with the physical conditions of life. It is the doctrine of Malthus applied with manifold force to the whole animal and vegetable kingdoms; for in this case there can be no artificial increase of food, and no prudential restraint from marriage. Although some species may be now increasing, more or less rapidly, in numbers, all cannot do so, for the world would not hold them.

Now, fairly obviously, nothing is very formal here. Premises or axioms are not stated explicitly; rules of inference are not given; and a conclusion about the struggle is not rigorously inferred. On the other hand, one might, perhaps, with some justice argue that a fairly rigorous argument is struggling (if the reader will forgive the pun) to get out, that with a minimum amount of *reconstruction* Darwin's informal argument can be put into a more formal mold. Thus, for example, his first sentence unequivocally shows that one of his premises is the high rate of organic tendency to increase. Without wanting to make claims that this is the only, or even the best, way of reconstructing Darwin's argument, one might produce something like this:

Premise I: Organic beings tend to increase at a high (geometrical) rate.
Premise II: If organic beings tend to increase at a high rate then either there must be a struggle for existence, or the numbers of organisms go up without limit.
Premise III: If the numbers of organisms go up without limit then the world must have unlimited room.
Premise IV: The world does not have unlimited room.
Conclusion: There is a struggle for existence.

5. Darwin's Appeal to Laws

Now, what about the lawlike nature of these claims? I have mentioned that laws are usually thought to be true, universal statements about the world. Laws are distinguished from other such statements because in some way they are thought necessary (a necessity often labelled "nomic" necessity). Thus, to take a famous example from optics, Snell's law tells us that there is a definite relationship between the angle of incidence (i) and the angle of refraction (r) when light is refracted upon passing from one medium to another. In particular:

$$\frac{\sin i}{\sin r} = R$$

Somehow, we feel that this *has* to hold, as that all my children are under ten, although true, does not. It could well be that some of my children are over ten, and, indeed, some day I hope this will be so.

Why do we feel that laws are necessary? Often, no doubt, because they are connected in axiom systems with other laws we think necessary. But ultimately, we probably feel this way because we have so much empirical evidence of the laws holding in so many different situations and no counter evidence; so somehow we feel the laws *must* hold. On the other hand, one has so much evidence of men of my age with over-ten-year-old children that no necessity is felt in my case. (I am here conveniently suppressing troublesome phenomena for Snell's law like Icelandic Spar. In my book, referred to at the end of this chapter, I take up this point.)

Turning to the reconstruction of Darwin's argument: One might fairly happily concede that, judged by the kind of criteria just discussed, the first premise is indeed a law. Organisms do seem generally to have such a tendency to increase—men as much as field mice. And, of course, in the long run it makes little difference how quickly or slowly organisms breed, although, of course, the limitations of room are liable to come into action much more rapidly for fast than for slow breeders. On the other hand, Premise IV is clearly not a law. It is not a universal statement: It makes an assertion about a property of one thing, the world.

What do we do at this point? Some philosophers have argued that *all* the empirical statements of theories of the physical sciences are laws, that such theories make no existential assumptions. In other words, physical-science theories are like sausage machines: They can process facts about individual things in the world if and when they are fed into the theories (in the form of "initial conditions"), but until such facts are indeed fed in, they merely make assertions about possible, hypothetical situations. Obviously, if this is the case, then more reconstruction of Darwin's theory is needed. One must, for example, replace Darwin's Premise IV with something like:

78

Premise IV': Any group of organisms increasing without limit will eventually take up more room than any arbitrary specified limit.

And corresponding changes must be made in the other premises.

Now perhaps we can begin to feel that we have whipped Darwin's argument into the kind of shape to be found in the physical sciences, at least with respect to structure and laws. The trouble is, of course, that in order to do so, we have had to engage in a moderately strenuous reconstructive programme. Do we, therefore, have any right to say that our reconstructed argument is *really* Darwin's argument? One is reminded at this point of a kind of word game that children like to play. One starts with one word and keeps changing a letter until all are changed, but stays with intelligible words: HAND, HIND, HINT, HILT, KILT.

One's model of a physical theory corresponds to HAND. Obviously if one's biological theory corresponded to KILT, it would not be the same. But what if it is HIND or HINT, which is about where Darwin's theory seems to come? Can we legitimately say that a reconstruction of HIND or HINT to HAND is permissible in order to claim that Darwin's argument is of the HAND kind? Perhaps one move would be to argue that physical theories rarely exactly exemplify the HAND type. They are often more akin to the HIND or HINT type. One might argue, for example, that physicists often get close to incorporating existential assumptions in their theories, as when the astronomer makes direct references to *our* sun and *our* moon. In this case, say, the HINT form of Darwin's theory is not necessarily so very different from the form of a physical theory.

I think that this is the kind of position I myself would want to take. Darwin's argument is not very formal; but it does seem to have a recognizable structure, and laws seem to have been appealed to. In sum, I think it not overly generous to suggest that the argument is not so very different from an argument of physics.

Of course, this argument, although important, is only a small part of Darwin's overall theory, and one might well feel a lot less sanguine about the structure of the rest. But before turning to the theory as a whole, we should just briefly consider a very common criticism of Darwin's mechanism of natural selection, which is all the more pressing because it is often made against the modern notion of selection too.

6. Natural Selection

Although physical theories make extensive use of logic and mathematics, we have seen that their distinctive mark is the inclusion of empirical laws of nature which distinguish them from other bodies of knowledge. Theories tell us about what happens in the universe, and although we may think them in some way necessarily true, ultimately the check of their truth and their usefulness depends on this reference to experience. Thus, for example, real bodies either attract each other with a force proportional to the inverse of the square of the distance between them, or they do not. But, it is argued, Darwinian evolutionary theory fails to meet such standards, and, in fact, fails to meet any legitimate scientific standards at all because its central concept, natural selection, in no way can be put in the form of a general law. It is vacuous or tautological. Hence, evolutionary theory is at best a fancy redescription of the obvious.

The nature of this objection is fairly simple to see, as also is its refutation. Natural selection states that certain organisms, by definition the "fittest," survive and reproduce. Hence, so the objection goes, that the fittest survive is merely saying that those which survive are those which survive, a statement which, although true, is hardly very informative and certainly not the basis of a full-blooded scientific theory. But there is far more to the concept than this, and as soon as it is fully understood one's worries vanish—or, at least, they ought to vanish. For a start, the concept points to a differential survival and reproduction in the world: Not all organisms that are born survive and reproduce, and there is certainly nothing very vacuous about this fact. It could be false (even though it is not): There might be a fixed number of non-breeding immortal organisms. Secondly, the concept also emphasizes that it is collections of certain peculiar characteristics which enable their possessors to survive and reproduce—survival and reproduction is not just a matter of chance, indifferent to the differences between organisms. Moreover, biologists believe that there is a certain consistency about these characteristics: Those which enable organisms to survive and reproduce in one set of circumstances will do the same in similar circumstances. All of these claims could be false. Hence, although there may indeed be great practical difficulties in deciding which are the pertinent useful (adaptive) characteristics, there is nothing particularly vacuous or tautological about the concept of natural selection itself.

7. The Overall Structure of Darwin's Theory

Take Darwin's theory as a whole: He applies the concept of natural selection to different phenomena—instinct, paleontology, geographical variation, embryology, morphology, and so on. Thus, for example, he argues that birds of different species on different islands of archipelagos often resemble each other because they are all descended from a common ancestor or group of ancestors. A small group, blown from the mainland, first to one island and from thence to others evolved through natural selection with different characteristics adapted to the different ecological niches of the various islands. It is interesting to note that historically it was the differences between the finches on the different islands of the Galapagos Archipelago that was perhaps the major factor in turning Darwin into a evolutionist. Similarly, Darwin argues that the embryos of quite different species are often very similar, despite great differences in the adult forms, because the embryos in the wombs have similar conditions, and thus it is only as these embryos start to grow up that selection drives them into their different forms.

To refer back to the criteria abstracted from the physical sciences: What can we then say about the overall theory of Darwin? I am afraid that if we look for even a sketchy overall hypothetico-deductive structure, things are not very hopeful. Take, for example, the birds on archipelagoes: Darwin gives a few examples, discusses how natural selection might act in such a situation, and so on. There is, however, nothing even remotely resembling a rigorous deductive argument. Apart from anything else, Darwin had no theory of heredity. He had no idea how favourable variations might be passed on from one generation to another and thus accumulate in an evolutionary fashion. With such a lacuna as this in his knowledge, he was perhaps wise not to attempt anything too rigorous. In short, we have rather loose links between Darwin's mechanism and the conclusions he would draw about different phenomena. From this together with our earlier conclusions, we might decide that some of Darwin's central arguments are hypothetico-deductive, but that overall the theory is at best a rather faint approximation to such a model.

What conclusion should we therefore draw—that the hypothetico-deductive model is irrelevant to evolutionary theory? That perhaps biological understanding does not share the norms of physics? Or that the theory of evolution as Darwin presented it was in essential respects rather immature, waiting for further development—development which would bring it closer to the ideal of physics? Although perhaps not necessarily, this does seem rather to be a question of history: If it can be shown that in the hundred years since the *Origin* evolutionists have made a significant effort to bring their theory closer to the physical ideal, probably one's feeling will be that Darwin's theory should be judged as rather immature. There is probably no need to seek alternative ideals for biology. Let us not, therefore, commit ourselves prematurely to answering this question, but instead hold our peace until the later version of evolutionary theory has been examined.

81

8. Consilience

But there are the two other criteria of physics to be considered—consilience and teleology. As consilience probably is the point at which Darwin's most resembles a physical theory like astronomy. Most striking about Darwin's theory is his attempt to apply his mechanism of natural selection in so many different areas of biology, as enumerated above. Moreover, we know that he did this consciously, modelling his theory on astronomy, and constantly used the consilience as a defence against critics. Furthermore, one feels that within the limitations set by the weaknesses of the links between his mechanism and his conclusions, he was highly successful in this and had every right to parade his success. As every reader knows, one of the most common criticisms of evolutionary theory is that "it is only a theory." Ignoring the obvious retort that "so is everything else in science," it is clear that the basic worry is that in some important way evolutionary theory is unproven. I would suggest that although we should not blind ourselves to what we might decide is the looseness of Darwin's theory, were we to concentrate a little more on the consilience, our feeling about the theory's basic, general truth and the mechanism of natural selection, in particular, might be much improved. Is it really reasonable to suppose false or unproven a theory which shows simultaneously why my hand is like a bat's wing, why baby humans look like baby dogs, why there are all kinds of different finches scattered over the Galapagos, why the fossil record is as it is? One could answer in the affirmative only if convinced that God is setting out systematically to deceive us in every aspect.

Since claims that evolutionary theory (Darwin's or the modern version) is essentially unproven are so common, let me proceed further with this point. Critics of evolutionary theory usually make the basic mistake of confusing the evidence for it with the evidence for phylogenies, where by this latter term is meant the actual paths that evolving organisms took. Although the fossil record gives us some very well documented cases of the evolution of some organisms—the evolution of the horse from eohippus is the classic example—it has to be admitted that often very little is known about the actual way in which organisms evolved. But the *theory* of evolution is concerned with the general reasons behind change, and thus does not stand or fall by our knowledge or ignorance of the fossil record. Physicists can separate speculations about the origin of the universe from Newtonian (or Einsteinian) speculations about the laws governing the universe: This surely is the right approach to take in evolutionary studies.

Darwin's way of supporting his evolutionary mechanism, a way which seems quite unchanged today, is very much like the tactics employed by a prosecutor in a trial who is trying to convict on circumstantial evidence. No one saw the butler murder Lord Rake; but we know that the butler's alibi was false, that the butler is an expert marksman, that Lord Rake seduced the butler's daughter, and so on. In short, the prosecutor argues that the butler's guilt is "beyond reasonable doubt." This is the claim that evolutionists from Darwin on make about their theory, and is, I think, a fair way of arguing.

9. Teleology

I come now to the question of teleology. Physicists do not talk in terms of "functions": The moon does not go in front of the sun "in order to" black out the earth. On the other hand, Darwin—here again, we can include modern evolutionists—quite cheerfully employed teleological language. Thus, it makes perfectly good sense to ask what function a woman's breasts serve: They exist in order to feed the very young and, no doubt, most biologists would also allow, in order to attract mates. Does this, therefore, mean that evolutionary biology, from Darwin to the present, is in some significant way different from a theory like Newtonian astronomy which does not use such teleological language?

To answer this question, let us begin by asking why evolutionists use function-talk. The answer, historically, is that they took it over, lock, stock, and barrel, from earlier nonevolutionists. Before Darwin, practically everyone believed that organisms are direct creations of God. In other words, organisms are God's artifacts, and just as we can use teleological language about our artifacts, so it is also appropriate to use teleological language about God's. When I put together two pieces of ground glass, I do it *in order that* I might better view the moon; the telescope serves the *function* of letting me look closely at the moon. Similarly, when God created the eye, He did it *in order that* we might see. It serves the *function* of sight. Darwin eliminated this direct reference to God—at best God creates and maintains through the remote control of unvarying laws. For the evolutionist, something like the eye is the product of natural selection: Evolving from rudimentary forms, those organisms with more efficient eyes had the edge in the battle to survive and reproduce. It is clear, therefore, that the evolutionist does not differ from the physicist in the most important matters. He does not believe that organisms are symbolic of some special creative intention, in a way that the objects of physics are not.

On the other hand, Darwin continued to talk quite cheerfully *as if* the eye had been specially created. And more generally, whenever one was faced with an advantageous characteristic, an adaptation, functional language was used. The metaphor of organisms as human artifacts was never eliminated from biology, and indeed, still continues strong as an explanatory and heuristic principle. When faced with some strange characteristic, the evolutionist's first question is, "What's it for?" Does this mean that in this respect, at least, biology is essentially different from physics? Probably, but this is not quite as significant as one might think. Because natural selection produces organisms which really do resemble artifacts—the heart is just like a man-made pump; the eye is just like a man-made telescope—biologists find it highly convenient to treat organisms as if they were artifacts. No such compulsion or consequent benefit is felt by physicists: The moon is just not much like an artifact. This does not, however, signify that biologists think that organisms really are artifacts, nor does it deny that biologists might drop function-talk from their science. For example, instead of speaking of the heart as existing for the function of blood-circulating, one might always speak of the heart as a product of a long, past process of natural selection. In this case the teleology appears to have vanished: One no longer thinks in terms of what a present heart *will* do.

Of course one might argue, and indeed I think some philosophers and biologists would argue, that functions only, in fact, signify past selection processes, and so there is nothing at all genuinely teleological about biological function-talk. My own feeling, however, is that in employing the artifact-metaphor and talking of functions, one is trying to understand organisms in terms of what one thinks they *will* do. Obviously, the reference to past selection-processes is still there—one argues to the future on the basis of what one thinks about the past— but I would suggest that to think that reference to the past is everything is slightly to truncate the biologist's full intention. In short, I think one could probably eliminate teleology from evolutionary thought without a great loss of content—although, I suspect, only by surrendering a very useful heuristic principle. But as things stand, I think the teleology of biology is there, and that in this respect biology differs from physics.

10. Mendelian Genetics

Let us turn from Darwin to the last century's evolutionary studies. Undoubtedly, the greatest advance has been to expand our understanding of the principles of heredity: the causes of characteristics and how they are transmitted from one generation to the next. Beginning with the seminal work of the Austrian monk Gregor Mendel, much is now known on this subject which has greatly augmented evolutionary theory since the days of the *Origin*. In particular, we now know that the units of heredity, the genes, are carried on string-like things, the chromosomes, in the cells of organisms. Furthermore, we know that the genes are passed on according to well-known rules, and it will be convenient here to examine the best-known of these: Mendel's first law of hereditary transmission.

To grasp the law, we must first understand that the chromosomes occur in pairs, and that consequently each gene (in a normal cell) has a corresponding gene which is said to be at the same "locus." The law then states: "For each sexual individual, each parent contributes one and only one of the genes at every locus. These genes come from the corresponding loci in the parents, and the chance of any parental gene being transmitted is the same as the chance of the other gene at the same parental locus." In other words, considering just genes at one locus, if the parents have genes AB and CD, then $1/4$ of the offspring will on average have AC, $1/4$ will have AD, $1/4$ BC, and $1/4$ BD. If the genes at a locus (called "alleles") are the same in an individual, the individual is said to be "homozygous," otherwise "heterozygous."

For evolutionary purposes it is necessary to generalize this law for group situations: It is the essence of any evolutionary theory founded on natural selection that it is not the individual *per se* which evolves, but the group. The generalization is fairly simply, and (after those who first produced it) is known as the Hardy-Weinberg law. For the law we need a large population of random-mating sexual organisms. Confining ourselves for simplicity to one locus, and supposing that there are only two alleles at this locus, A_1 and A_2, thus giving rise to individuals of types, A_1A_1, A_1A_2, and A_2A_2, let us suppose that the initial ratio of A_1 to A_2 genes is $p:q$. Then, barring any outside influences, the Hardy-Weinberg law states that for all generations following, the ratio will stay at $p:q$, distributed amongst organisms as follows:

$$p^2 A_1A_1 : 2\,pq\,A_1A_2 : q^2\,A_2A_2$$

The Hardy-Weinberg law is the foundation of modern evolutionary thought. Essentially, it tells us that gene ratios (and hence physical-characteristics ratios) will remain just about unchanged, unless something comes in to alter them. This may seem a bit unexciting, but its role, seems essentially akin to that of Newton's first law of motion which enabled the physicist to proceed from a background of stability. He knows that little balls do not just speed up or slow down at will—any change means a definite cause. Similarly, proceeding from the Hardy-Weinberg law the evolutionist can introduce quantified causes of change knowing that they will not be swamped by an unpredictable background of flux.

85

The evolutionist believes there are two main causes of change: First, our old friend *selection*, very similar to Darwin's except that the modern evolutionist usually thinks in terms of genes rather than individuals. The other major cause of change is *mutation*. It is believed that sometimes the genes change spontaneously: This causes new variation which—given selection—leads to the evolution of new forms of organism. Mutation is believed "random" because its occurrence is independent of the needs of an organism, but its rate does fortunately seem to be quantifiable.

This theory of Mendelian genetics (whose generalized form is usually called "population genetics") is then applied to different areas of study—geographical variation, for example—very much like Darwin's speculations were applied. Much of what I have already said, therefore, can be applied without further comment from Darwin's theory to modern theory. In particular, no more need be said here about the third and fourth criteria of physics, consilience and lack of teleology. In modern evolutionary theory we find a consilience and the use of teleological language. Let us, therefore, turn to our first two points, the structure of modern theory and the putative lawlike nature of its claims.

11. The Structure of Modern Evolutionary Theory

It is undeniable that, in its structure, the populational-genetical part of modern evolutionary theory comes much closer to being a rigorous deductive system than anything found in the *Origin*. To take but one example: Sickle-cell anaemia, a form found in certain African tribes and almost invariably fatal in childhood, is caused by a mutant gene— homozygotes for the gene having the anaemia. It is also known, however, that heterozygotes for the gene, far from dying from anaemia, have a genetic immunity to malaria—no small boon in a place like the Gold Coast. Starting with the Hardy-Weinberg law, it is possible to show rigorously how, as it were, it is selectively advantageous for a few to be sacrificed to anaemia for the sake of increased malarial immunity—a situation which can and does continue indefinitely. (Full details of this case are given in my book.)

The axiomatic ideal is, therefore, much better exemplified in modern evolutionary theory than it is in the *Origin*. The modern theory is, thus, closer to being like a theory of physics. But the whole theory is not now rigorously deductive—far from it. When population genetics is applied to phenomena like the distribution of organisms on oceanic islands, a great many guesses have to be made and gaps conceded. We can go much farther than Darwin, but by no means all the way.

Because we are now closer to satisfying the axiomatic ideal, I would, nevertheless, now answer a question earlier left dangling. Since evolutionists in the past century have brought their theory much closer to the ideal of physics, I would suggest that it is reasonable to conclude that they do share the ideals of physics in this respect, and that their theories be judged immature rather than fundamentally different. If evolutionists had shown themselves quite indifferent to the axiomatic ideal, then perhaps a case could be made for the peculiarity of biological understanding. But since their work seems to get ever closer to the ideal, surely it is indeed their ideal.

Finally, what about the lawlike nature of the claims of population genetics? Are Mendel's first law and its generalization, the Hardy-Weinberg law, genuine laws? I think they are: Mendel's first law holds for organisms as divergent as pea-plants and men. The evidence is so overwhelming that one feels that it *must* hold, and since the Hardy-Weinberg law is its consequence, one feels the same about it too. Long after Mendel published his laws citing evidence based on pea-plants, it was found that his experimental results were far too good for real life. So apparently Mendel was convinced of the necessity of his laws even before he had any evidence!

In short, therefore, I would argue that modern evolutionary theory is in many respects like a theory of physics. In the sense already discussed, however, it seems rather immature, and because organisms look like artifacts, it has a non-physical teleological element.

12. Are Physics and Chemistry Taking Over?

If, as I have suggested, a major biological theory is in important respects like a physical theory, and if, as I have also suggested, in the past century it has come even closer to being like a physical theory, should we not then look to the day when it will in fact become a physical theory? And more generally, will biology some day be made part of physics?

There is no logical connexion between a biological theory being like a physical theory, and its becoming one. It could ultimately prove that the organic and the inorganic world are irreducibly different. But, as the reader will know, in the past twenty-five years moves have been made which make the take-over of biology by physics (or physics and chemistry) seem far less abstract. In particular, it is now recognized that genetic information, something which was a paradigmatically biological phenomenon if anything was, can be traced ultimately to long molecules of nucleic acid (DNA and RNA), and it is known that information crucial to heredity is a direct function of the physico-chemical composition of those molecules. In other words, it really does seem that something which hitherto was exclusively biological is now being shown as a consequence of organisms considered as physico-chemical objects.

In reply to the suggestion that perhaps biology is now becoming part of physics and chemistry, I would want to make two comments: However much molecularization of biology is going to take place in our lifetimes, it would seem a bit ridiculous to conclude that biology as an autonomous discipline will vanish entirely. Without in any way denying the great influence of so-called molecular biology, we are a long way from being able to deduce, for example, the sex life of the fruit fly from the basic principles of quantum mechanics. There is still—and probably in practice always will be—a very important place for the study of organisms considered as *biological* objects.

Secondly, however, I must concede that I do see a very significant link between biology and physics, and problems which were once biological like those of genetics, falling to the province of physics and chemistry. Of course, in one sense organisms have always been susceptible to a physico-chemical analysis: A falling elephant obeys Galileo's laws no less than a falling rock. But now it seems that *biological* properties of organisms are being explained by physico-chemical principles.

The exact relationship between the new physico-chemical and the old biological theories is a matter of considerable debate. Some philosophers see the relationship as one of *reduction*: Biological theories are shown in some way to be deductive consequences of physical theories. Others see the relationship as one of *replacement*: The new physical theories show the old biological theories to be false, and thus firmly and not altogether politely push them to one side. Without entering into details of the controversy here, let me say that I incline to the former position, for I really do see the molecular theory about DNA and so on showing precisely why biological laws like Mendel's hold. (My reluctance to fight here is solely a function of limited space; in my book I do put on the gloves.) But whatever the relationship may be between the new molecular and the older biology, the significance of the former is undeniable.

13. The Autonomy of Biology

Are there in principle any limits to molecular biology's influence? A number of biologists and philosophers think there are aspects of the biological world which in principle could never be adequately explained by the physical sciences. I have myself already veered close to this position in my discussion of teleology by allowing an irreducibly teleological element in evolutionary biology. We could eliminate it perhaps, but then our understanding that organisms seem so very much like artifacts would be lost. One might feel that the benefits of a physico-chemical approach would compensate for this, but I think that there would be a definite loss. Unless, of course, one brought a little teleology into physics and chemistry! This is not as crazy an idea as it sounds: One would not be bringing teleology into physics and chemistry until physics and chemistry were already considering aspects of organisms that pertained to their artifact nature—not, in other words, until physics and chemistry dealt with the very aspects of organisms which make biologists resort to teleological understanding. And, indeed, it is worth noting that already molecular biologists use artifact-type language, as, for example, when they refer to the structure of the DNA molecule as written in a genetic "code."

The defenders of biological autonomy, however, usually want a stronger defence than mine. They want to argue that in very fundamental ways biology must always remain a subject apart. Unfortunately, I suspect that although I can defend my position, they cannot defend theirs. Their most popular argument—and the only one I shall consider here—is that biological phenomena are in some important way *unique*. Physics and chemistry, it is argued, deal with the repeatable, like molecules and planets. Biology deals with the unique: individual organisms and the evolution of non-repeatable species like man. Biology, therefore, needs, and always will need, a mode of understanding different from the physico-chemical. But it is easy to show, that this argument rests on an equivocation which, when once exposed, causes everything to collapse. In one sense, everything is unique; in another, nothing is. Although the evolution of the African elephant may be unique, the evolution of elephants is not, for there are Indian elephants. Conversely, although the elliptical paths of planets are not unique, the particular path of Mars is unique. By increasing the defining properties, we can make anything unique; by relaxing the number of properties specified, almost anything becomes non-unique. The trick in science is to keep the number of specified properties so sufficiently low that we can keep the number of objects specified high enough to be of interest, whilst we keep both the number and kind of properties high enough to make what we say interesting.

But biologists do, of course, manage in practice to find repeatability. Mendelian genes, for example, are supposed to occur over and over again in different organisms, and just as one molecule of water is supposed identical to the next, so also one copy of a particular gene is supposed absolutely identical to the next. In short, the objects of biology seem no more—or less—unique than the objects of physics. The autonomy of biology cannot be preserved in this way.

Selected Bibliography

Biology

Charles Darwin, *On the Origin of Species*. There are many reprinted editions of this work. I would suggest that published by Penguin—an inexpensive reprint of the first edition, much the best.

Ernst Mayr, *Animal Species and Evolution*. (1963) Cambridge, Mass: Belknap. Th. Dobzhansky, *Genetics of the Evolutionary Process*. (1970) New York: Columbia University Press. These are two highly readable and authoritative introductions to modern evolutionary thought.

Philosophy

Michael Ruse, *The Philosophy of Biology*. (1973) London: Hutchinson. David Hull, *Philosophy of Biological Science*. (1974) Englewood Cliffs: Prentice Hall. Both of these refer to and discuss most of the debates in philosophy of biology today.

Ronald Munson, *Man and Nature*. (1971) New York: Delta. The best collection of readings in the philosophy of biology, although already a little dated.

Neodarwinism, Mental Evolution, and the Mind-Body Problem

THOMAS A. GOUDGE

I

The scientific theory of evolution founded by Darwin in 1859 and developed into Neodarwinism or the "synthetic theory" in the twentieth century has aspects of great philosophical interest. One is the bearing of the theory on questions about mental powers or minds. For these powers are clearly characteristics of some living things, and hence, it would seem, must have undergone a lengthy process of evolution, like all else in the organic world. Darwin dealt with this subject in several chapters of *The Descent of Man* (1871), a sequel to his major work, *The Origin of Species*. One would have thought that it was but a short step to the discussion of the classical problem of the relation between mind and body.

Yet for the most part, evolutionary ideas have had very little impact on philosophical treatments of this classical problem. Part of the reason for the minimal impact is, no doubt, that the basic terms of the problem had been established long before Darwin came on the scene. A number of theories had been advanced about how the human mind is to be conceived, how it is related to the physical body, and how it performs its different functions. But these matters remained profoundly controversial. Recently a new formulation, the "Identity Theory," has attracted wide attention among philosophers. But it, too, seems to be yielding diminishing returns. Thus, it looks as if some fresh approach to these complex issues may be needed.

What I wish to explore in the present paper is an approach made possible by evolutionary theory. My strategy is to shift discussion away from classical mind-body theories, with their exclusive emphasis on human mental powers, and direct it to the question of the place of those powers in the economy of nature, considering how they came to be what they are as a result of the process of organic evolution. This strategy will put the idea of mental evolution in the center of the stage. It will also involve reviewing a number of the details of Neodarwinian theory as well as certain rational conjectures based on it.

One other facet of the general strategy can be brought out by a reference to Cartesianism. Descartes, as has been said, still casts a long shadow over this whole subject. It is not just that many are explicitly influenced by his radical dualism of minds as immaterial substances and bodies as material automata, but also that sciences such as neurophysiology and neuroanatomy often implicitly assume that they are working on one side of the Cartesian bifurcation, since they do not profess to be able to explain how brain-processes are related to "states of mind." Rationality is often regarded as uniquely characterizing man, a typical Cartesian view, and so on. Even physicalism and central state materialism are theories based on one half of the body-mind bifurcation, and derive from it at least some of the effective contrast which delimits their meaning and gives them an opponent. One of the features of the evolutionary approach is that it shifts the context of the discussion from Descartes to Darwin and thereby seeks to by-pass some of the thinking which has led to the most intractable issues associated with Cartesianism.

One way to represent the proposed shift of context is to say that it replaces a special creationist view of mind by an evolutionary one. This evolutionary view is based on the facts and interpretations provided by the general theory of biological evolution, as Darwin made abundantly clear in the first part of *The Descent of Man*. Since the publication of that work, however, two developments of special significance for the topic of mental evolution have occurred: (*a*) There has been a change in the formulation of that part of the theory of evolution which specifies the causal factors which have determined in large measure what has occurred in the history of living things. Not only has the genetic basis of variations, including mutations, been made clear, but so also has the statistical or populational character of natural selection, which is no longer construed as a "struggle for life" of individual organisms with each other and a simple "survival of the fittest." "Natural selection," indeed, is now taken to be a common name for several related, yet distinct, processes which go on in populations, and which eventuate in either increase or decrease of adaptation and adaptability of those populations. (*b*) There has been a recognition that man, and especially homo sapiens, has arisen in evolution primarily as a result of conditions which he himself has produced in the various types of culture. Other animals have evolved in consequence of genetic and ecological pressures of a biophysical kind. Man, too, has been subject to those same pressures; but far exceeding them in importance have been the ecological pressures of a cultural kind. In fact, one can strongly claim that man's cultural environment is his unique ecological niche, to which he is constantly seeking to adapt and which he is constantly seeking to change. These two developments—the reformulation of the Darwinian theory of evolution and the recognition of the distinction between genetic and cultural evolution—are of special significance for the understanding of mental evolution and its bearing on the body-mind problem.

II

Against this background, I shall consider three questions:

1. What is the meaning of the expression "mental evolution?"
2. What function or functions can be plausibly attributed to mental phenomena in the lives of individual animals and of populations during evolutionary history?
3. What consequences can be derived from the examination of these two questions for the understanding of mental phenomena in homo sapiens?

In order to prejudge as few matters as possible, I shall treat the expression "mental phenomena" as referring to a large spectrum of items such as are specified by the words "thinks," "understands," "feels," "decides," "speaks," "writes," "learns," "solves a problem," etc. The spectrum thus includes what in a dualistic terminology would be called "psychical" and "behavioural" phenomena. Some may object that treating "mental phenomena" as such a catch-all expression begs too many questions to be an acceptable basis for discussion. In reply, I would urge that I am introducing the usage provisionally, not apodictically; that it is in line with prephilosophical ways of speaking; and that a judgment about its soundness or unsoundness can be better made at the end rather than at the outset of the inquiry.

What Is Meant by "Mental Evolution"?

There are at least two approaches to this question: One consists of general arguments for the occurrence of mental evolution, the other in reconstructing the course of mental evolution among the successive forms of animals on the earth. I shall consider each briefly.

A general argument for mental evolution takes as a premiss that a long, complex process of organic evolution has occurred on the planet resulting in immense changes in living things. Among the changes are trends marked by a cumulative increase in the complexity and variety of living things and their phenotypic characteristics. Hence, mental phenomena have undergone changes from simple to complex, and from homogeneous to heterogenous forms. Such changes constitute an essential feature of mental evolution.

An objection to this argument might be that although it admits a distinction between somatic and mental phenomena, it *assumes* that they are sufficiently alike to conclude that the same evolutionary trends are exemplified by both. But this assumption can and will be challenged by a Cartesian, for example, or by a supporter of a special creationist view of the soul. Thus, we might agree that the somatic

93

characteristics of living things have evolved, and at the same time deny that mental characteristics have.

A reply to this objection is a second general argument for mental evolution which traces the consequences of the neurosciences (neuroanatomy, neurophysiology, neuropathology) showing that in higher animals mental phenomena are intimately connected with the presence and normal functioning of sense organs, nervous systems and brains in those animals. Indeed, the connection is so intimate that there are strong reasons for holding that mental phenomena cannot exist in the absence of the functioning of sense organs, nervous systems, and brains. Now these somatic structures have had a long evolutionary history which it is possible to reconstruct at least in outline. The reconstruction shows that the history has been a broadly continuous, cumulative progression in which sense organs, nerves, and brains first appear as very simple, unspecialized structures and gradually become more and more complex, diverse, and specialized in different types of animals. Comparative studies of animals now on the earth support the view that complexity of functioning accompanies complexity of structure. Accordingly, the phylogenetic development of neural systems points unmistakably to the conclusion that there has been a phylogenetic development of mental phenomena which are existentially dependent on those systems.

A more detailed formulation of this argument would have to introduce various technicalities which I must by-pass, save for one point that possesses considerable intrinsic interest and does, I think, add force to the argument: a recent discovery in neuroscience that a sub-cortical region of the brain, known as "the brain-stem arousal system," plays an important part in determining mental activities in humans. While the matter is not entirely beyond dispute, there are reasons for believing that this sub-cortical region, and not the cerebral cortex alone, is required to account for some human intellectual processes. Now the brain-stem region is phylogenetically older than the cortical region and is a more "primitive" structure which man shares with many other animals who appeared before he did on the evolutionary stage. If so, this discovery adds support to the conclusion that mental phenomena occur in animals whose central nervous systems are a good deal simpler than the human central nervous system, and that such mental phenomena are also simpler than those found in normal human beings.

Only a few attempts to reconstruct the main stages of mental evolution have been made, partly, no doubt, because so much of the enterprise has at present to be conjectural. One of the most impressive attempts is that of the zoologist Rensch who suggests that in the lowest multicellular invertebrates possessing neurons and diffusely distributed nerve-nets (e.g., jellyfish, sea-anemones, starfish, sea-urchins, etc., which originated in the Cambrian period), one may suppose isolated sensations to occur. These would be among the earliest items of the spectrum of mental phenomena. Somewhat more determinate would be temporary associated sensations in lower invertebrates whose neuronic systems allow synaptic connections. Then, perhaps, came the beginning of longer lasting coherence of awareness by correspondingly longer, spontaneous activity of neurons in the beginnings of a central nervous system (e.g., spontaneous electrical activity which begins in hydroid polyps). The beginnings of a brain can be seen in certain flatworms (planaria), and so from about this point on, the progressive complication and diversification in different animal forms of the central nervous system and the brain can be traced in outline. Thus, the process can be traced through the lower vertebrates such as the bony fishes (Osteichthyes) and insects, to the birds, the mammals, and to that subdivision of the mammals, the primates (who appeared in the Eocene or possibly the Paleocene, c. 70 to 80 million years ago). With reference to the later forms, whose descendants are being increasingly studied by ethologists and primatologists, a growing body of evidence permits informed conjectures to be made about the stages in mental evolution when such things as types of learning, elementary "averbal concepts" (Rensch), aesthetic preferences, etc. began to appear.

Man's immediate ancestors closely resembled present-day apes and the line of descent passed from these primates through the "man-apes" (Australopithecines or some collateral group). Hence it may be inferred from the study of present-day apes that antecedents of most, if not all, man's mental processes existed in a less developed form in these ancestral beings. Rensch summarizes this facet of the argument thus:

> The brain structure of monkeys, especially of apes, in morphological, histological, and electrophysiological respects, is very similar to man (and even more similar to man's ancestors in the *Australopithecus* group). Only the frontal and temporal lobes are less differentiated, and the motor area of speech is lacking. The sense organs of apes and monkeys are fundamentally equal to those of man. These facts, and the rather well-established and theoretically continuous line of descent to humans strongly suggest that apes and monkeys must be credited with psychical processes.

Moreover, between the period of the Australopithecines (roughly, 1,000,000 years ago) and the appearance of homo sapiens (roughly, 250,000 years ago) the brain increased in size, and presumably in complexity of functioning, at a *very rapid* rate. It was probably near the beginning of his evolutionary appearance that homo sapiens (and perhaps a few closely related species) gained the vitally important motor area of speech (Broca's region) and thus initiated the evolution of human language.

Before leaving this topic, it is worth reminding ourselves that the reconstruction of the main stages of mental evolution will be misleading if it suggests the image of a smooth, continuous, linear sequence, or a neatly constructed "ladder" from "lowest" to "highest." Such a simplistic image is quite mistaken. Both somatic and mental evolution have indeed involved a *net, overall increase* in complexity, diversity, and variety of their respective items. But this increase has occurred erratically, at widely differing rates, with occasional stoppage and even reverses along the way, and probably also, with occasional quantum changes at critical points. Neither somatic nor mental evolution can be correctly construed as "perfecting" processes, since their cumulative character results not only in the retention of beneficial structures and functions but also in the retention of some that are not beneficial, given changed circumstances. The bearing of this last point on certain aspects of mental evolution may call for comment later in the paper.

What Functions Have Mental Phenomena Had in Evolution?

In approaching this second question, I shall start by recalling that I have taken the expression "mental phenomena" to refer to a large spectrum of items which, zoologically speaking, belong to the still larger class of phenotypic traits of animals. Now phenotypic traits are the joint product of an animal's genetic system or genotype, and the environment in which it is viable. According to the tenets of Neodarwinian theory, phenotypic traits are being constantly screened by natural selection with the result that those which are adaptively valuable to their possessors endure and undergo evolution, while those which lack adaptive value tend to be eliminated and at the same time, their possessors tend to become extinct. Using this consequence of Neodarwinian theory as a clue, it is reasonable to infer that mental phenomena have had an *adaptive role or function in evolution*, just as lungs and limbs have had. At any rate, I am going to take this inference as a hypothetical answer to the second question and investigate a few of its details.

But if mental phenomena have had an adaptive function in evolution, they cannot be devoid of causal efficacy, i.e., they cannot be mere epiphenomena. Moreover, the causality involved is not adequately envisaged as a simple dyadic relation. It is rather a causal network involving multiple relations, and perhaps best described in cybernetic terms, such as those of information flow, feed-back, storage, self-regulation, etc. The resources of cybernetic description and explanation in this area are only beginning to be drawn on and may eventually yield significant returns.

With this recognition of the need to assign a place to mental phenomena in the causal network of relevant events, if we regard the phenomena as adaptive, let us see whether we can make out something more of the meaning of "adaptive" in this particular context.

We may start by noting one important biological criterion of adaptation, viz., survival and reproduction. If an individual plant or animal is adapted to a particular habitat, then in general it lives to maturity in that habitat. If the population of which it is a member is adapted, then the reproductive rate must be such that its numbers do not diminish in a few generations to zero, or rapidly approximate to zero. Hence the majority of adapted members of the population must not only be able to survive to maturity but must also be fertile, and thus ensure that the numbers of the population will increase, or remain relatively stable, or at least diminish slowly. This criterion is relevant to cases where the habitat is broadly unchanging. If it starts to change substantially, the viability of individuals and populations depends not on their adaptation but on their adaptability. Adaptability is conditioned by mutation rates which ensure a supply of favourable mutations at that juncture of the population's history. Note, incidentally, that what goes on in this whole situation is to be interpreted statistically. Adaptation is never, or is very rarely, an "all or nothing" affair.

To what extent are mental phenomena adaptive in the above sense? It is not difficult to understand how they are, or can be, so. The possession of simple neuronic systems increases the amount of information available to the animals that have them, and so can make their responses to the immediate environment more effective by facilitating escape from predators and the locating of mates. The evolution of distance receptors, such as vertebrate eyes and ears, conferred obvious adaptive benefits. Similarly, evolution of proprioceptors by which animals obtain information about the movement of their limbs, the position of their body, etc., enhanced their capacity to cope with the world. The biological role of feelings and emotions was well documented by Darwin in his classical study of this subject. In short, as Julian Huxley has put it:

> The intensifying of the organization of the brains of animals, combined with the information received from the sense-organs and operating through the machinery of interconnected neurons, is of advantage for the simple reason that it gives a fuller awareness of both outer and inner situations; it therefore provides better guidance for behaviour . . . in the

97

complexity of the situations with which animal organisms can be confronted. It endows the organism with better operational efficiency.

While all this can be granted, little reflection is needed to convince one that if an adequate account of the function of mental phenomena in homo sapiens is to be forthcoming in the above terms, then "adaptation" will have to be understood in a *wider* sense than the exclusively biological. For as I have previously mentioned, the conditions of life of our species are predominantly cultural and not just biophysical. Hence we will have to try expanding the meaning of "adaptation" (and also of "environment") so as to construe at least some mental phenomena in man as ways of adapting, or seeking to adapt, to his *cultural* habitat.

If we make such a move, we can cope with a difficulty which faced several prominent Darwinians in the nineteenth century. Alfred Russel Wallace, G.J. Romanes, Asa Gray, and others, although accepting in general Darwin's theory as an explanation of how biological organisms, including the human body, had evolved, rejected the idea that the human mind has resulted from evolution by natural selection. Their rejection sprang from what they considered to be the impossibility of attaching any adaptive value to mental processes such as aesthetic enjoyment, mathematical thinking, religious experience, moral decisions, etc. These "higher" mental operations have no direct relation to man's biological requirements. They cannot contribute to success in the struggle for existence or help ensure survival of the fittest. Therefore, these Darwinians concluded, the human mind must lie beyond the explanatory limits of the theory advanced in *The Origin of Species* and *The Descent of Man*. Most of those who have discussed issues in the philosophy of mind since the time of Wallace, Romanes, and Gray have tacitly adopted the same position.

Now if the above difficulty can be removed in the way I am suggesting, by enlarging the concepts of adaptation and environment to include human cultures, a number of problems come up, only a few of which can be treated (and these only briefly) in a paper of this length.

First, to enlarge the concept of adaptation without generating a different concept, we need to retain some central features which attach to it at the biological level. Two such features, I suggest, are: (*a*) the connotation of *the usefulness* of biologically adaptive characteristics to individual organisms and populations, i.e., the capacity of those characteristics to lead to certain ends or goals. This was a matter that Darwin made clear by the doctrine of natural selection, which re-established teleology in biological thought by providing it with a scientific basis. (*b*) The connotation of the *harmonious fitting together* or *concordance* of organisms and their respective environments: A characteristic is adaptive in the degree to which it enables an organism to conform to its habitat or niche and simultaneously allows the habitat to meet the needs of the organism. I do not suggest for a moment that these two features exhaust the meaning of the concept of adaptation, which is complex and difficult to specify with precision. But I think that the two features can be retained in extending the concept to take account of man's relations to his cultural environment.

To effect a corresponding enlargement of the concept of the environment we need to recognize certain aspects of the term "culture" as understood by anthropology. Various elaborate definitions have been put forward, but the essential elements have been simply specified by Oakley: "The sum-total of what a particular human society practices, produces and thinks may be called its culture." As this statement implies, culture is dependent on homo sapiens as a species of social animal, whose survival requires him to live in groups. The main components of a culture, as Oakley points out are: (*a*) group practices and institutionalized forms of behaviour; (*b*) artifacts, i.e., economic products, art objects, technological items, etc.; and (*c*) languages, ideas, beliefs, theories, bodies of knowledge, i.e., intellectual products in the broad sense. At different stages in the evolution of cultures, one or other of these components may predominate over the rest, but in general they interact in complex ways. Moreover, a culture is not simply an aggregate, but is an organized *whole* or *system* with autonomy and objectivity, just like the biophysical environment. For at least the last 100,000 years, every member of the species homo sapiens has been born into a particular culture, about which and from which he learns as a child, to which at the outset he conforms, and with which he continuously interacts. Thus, each individual's culture as a whole is a milieu which calls upon his adaptive powers in much the same way as his biophysical environment does. But the adaptive powers called forth by the cultural environment are predominantly *mental.* Accordingly, in producing culture, homo sapiens brought into being a "third world" (to use Popper's phrase) by adapting to which a manifold and rapid development of his brain system and mental phenomena was engendered. In fact, one can say without too much exaggeration that human evolution has been predominantly mental evolution. These considerations seem to offer an interpretation of man's higher mental activities as adaptive, and also to have some significant implications for an evolutionary understanding of the body-mind problem.

As a test case for this way of construing man's mental activities, I will take up briefly Romanes' and Wallace's difficulty that at least some mental activities—for example, the ability to do advanced mathematics—cannot be plausibly accounted for in Darwinian terms. Now, it is certainly not plausible to attempt to connect this ability with biological needs of survival and repʌoduction. But it *is* plausible, I think, to relate it to certain cultural conditions. For mathematics, as a part of culture, is a human production and the body of knowledge it embraces is an autonomous sub-environment or domain within the cultural "third world." Those who learn about this domain and how to cope with it acquire an ability which is a specialized kind of adaptation. This ability enables those individuals to achieve certain purposes or ends, viz., the solving of problems, even highly abstract ones, with their science. The individuals "know their way about" (to adapt Wittgenstein's phrase), and "feel at home" in the mathematical environment. Moreover, some individuals create new features of that environment, often so substantial that the direction of its evolution is significantly altered. This fact is in line with the point previously mentioned—that adaptation is often a process by which living things act on the environment so as to change it, as well as being acted on by it. Likewise, differences in mathematical expertise square with adaptive functions having varying degrees from strong to weak. Let me add that I am not suggesting that the ability to do advanced mathematics is *only* an adaptive function. I am merely suggesting that a full account of what is involved in mathematical thinking must include its adaptive dimension. Hence, that account will not lie completely outside of, or be incompatible with, an evolutionary view of the matter. Furthermore, to contend that the ability to do advanced mathematics is culturally adaptive in no way conflicts with the contention that *rudimentary* mathematical ability is biologically useful, and so was favoured by natural selection from an early evolutionary stage. This is supported by the well-known and fascinating experiments of the ethologist Otto Koehler and his pupils on "counting" and "number recognition" in animals, especially birds.

This general line of argument, with suitable amendments to fit particular cases, can be extended to a wide variety of mental phenomena which might seem, prima facie, to be wholly detached from any evolutionary base. Moreover, in addition to the principles of interpretation already used, at least one other is available for such an extension of the argument. It is the principle that if several phenotypic characters are controlled by a single gene or constellation of genes, one of the characters may be non-adaptive, or even slightly maladaptive, provided that the other characters linked to it are markedly adaptive, for then the total set of characters will tend to be preserved by natural selection. By applying this principle to the cultural level, one might expect to find that some mental phenomena in humans, for example, may be non-adaptive and yet be preserved because they are culturally or psychologically linked to other mental phenomena which do have marked adaptive value. Human aesthetic enjoyment, for instance, may be one of these non-adaptive phenomena. I hardly need to add that much of all this needs further elaboration which cannot be given here. And it should not be forgotten that the additional principle just stated allows for another consequence, namely, that if a particular phenomenon linked to others in a complex becomes so maladaptive as to outweigh the combined advantages of those others, selective pressure will work against the whole complex. In that case, the end result can be the elimination of the complex as well as of organisms whose existence depends on it.

To conclude this part of the paper I will take up one objection to the line of argument just advanced. If, it may be objected, the prime function of mental phenomena is to promote the adaptation and adaptability of the human animal to its world, then one would expect the function to be manifested uniformly throughout the species. For in the case of a somatic characteristic, it would not be called "adaptive" if it had markedly harmful effects on large numbers of a population. Now the mental qualities of homo sapiens, particularly in the last 5,000 years, have undeniably had such harmful effects. One has only to think of the dismal record of wars, persecutions, enslavements, torture, exploitation, etc., which have resulted from man's proneness to adopt false beliefs, superstitions, and myths, reinforced by his violent passions and by outmoded forms of social organization, to recognize that he is the most destructive of all the animals. As the very survival of the species is threatened, it is surely more plausible to conclude that mentality, rather than being adaptive, is profoundly maladaptive.

To this objection, the reply can be made that just as somatic characteristics are rarely, if ever, adaptive in an absolutely universal or unconditional sense, but only in a statistical or "by and large" sense, so mental phenomena qua adaptive should be treated in the same way. Hence, alongside the record of man's inhumanity to his own and other species has to be put the record of his activities in promoting the welfare of both, through, for example, scientific medicine, conservation projects, the spread of general education, the creation of the arts, etc. Thus, man's destructive proclivities are opposed by constructive ones originating in his mental powers and directed towards enhancing his adaptability to the biophysical and cultural environments. Not the least of these powers is his intelligence and its cultural correlate, objective knowledge.

Some Consequences for the Mind-Body Problem

The strict Cartesian position, that mind is a psychical entity exclusively possessed by man, located inside his body-machine and interacting with it, and the correlative position, that animals have no mental processes of any kind, are untenable in the face of the facts and interpretations of evolutionary science, quite apart from other objections that can be made to those positions. This much seems to me quite certain, and by now is unlikely to be disputed by anyone familiar with the facts and interpretations.

The rejecting of Cartesianism implies rejecting not the existence of mental phenomena, but only the claim that these phenomena are either wholly or partly generated by the mind as psychical entity. Nor does the rejecting of Cartesianism imply that mental phenomena are devoid of causal efficacy. On the contrary, if, as I have argued, these phenomena have had, and continue to have adaptive value in general, as a result of the operation of natural selection in the pre-human stages of their evolution, then the phenomena must have become built into a causal network which operates in the animal-environment complex. Hence, the two traditional positions of epiphenomenalism and psychophysical parallelism are biologically implausible.

In the light of an evolutionary approach, one has to conclude that mental phenomena have had a long, broadly continuous development from relatively simple and elementary forms to more and more complex ones. As the complexity increased, various stages of organization of the phenomena can be distinguished in phase with various stages of complexity of neural systems. Moreover, in the case of each adult human being, and probably of other placental mammals, prenatal mental phenomena of a rudimentary sort are gradually developed in the course of embryogenesis, although there are no grounds for supposing the phenomena to be present at all in the *early* stages of the process. Ontogenesis in bisexual animals begins with the fusion of gametes in a zygote at fertilization. But there is no evidence, direct or indirect, that these cells have even generalized "awareness" or "consciousness," any more than they have endocrine activity or limb-movements, since the requisite anatomical structures are absent. J.B.S. Haldane summed up this situation as follows:

> The strangest thing about the origin of consciousness from unconsciousness is not that it happened once in the remote past, but that it happens in the life of every one of us. An early human embryo without nervous system or sense-organs, and no occupation but growth, has no more claim to consciousness than a plant—far less than a jellyfish. A new-born baby must be conscious, but has less title to rationality than a dog or ape. The

evolutionist makes the very modest claim that an increase in rationality such as every normal child shows in its lifetime has occurred in the ancestors of the human race in the last few million years.

Now in view of the broadly continuous character of phylogenesis and of ontogenesis, I think it becomes implausible to describe any mental phenomena as "psychical" in a sense of that term which implies a radical contrast with what is "physical." For in neither phylogenesis nor ontogenesis can any basis be found for a claim that something as radically different as psychical items are from physical items in Cartesianism entered at some historical point into phylogenetic or ontogenetic processes. But since, in discussions in philosophy of mind, "physical" draws much of its effective meaning from contrast with "psychical," I think we would be well-advised to avoid assertions such as "all mental phenomena are physical," or "a mental process is identical with some physical process in the central nervous system." What I have just said in no way obliterates the contrast between mental phenomena and neurological phenomena as different in various respects, but the difference is no more radical than the difference between, say, the endocrine glands and the stomach (or between hormones and pepsin). To put it otherwise (and perhaps pretentiously), we might well be guided in our attempts to unravel the "world-knot" of the body-mind problem by the view which evolutionary science suggests: that mental phenomena are *ontologically homogeneous with the other phenomena in nature* and causally meshed with those other phenomena, while at the same time mental phenomena have their own distinctive traits which have to be recognized.

III

Homo sapiens is the one species of animal which has been able to articulate a number of perplexities associated with the investigation and description of mental phenomena. Two which arise in connection with homo sapiens' own case have been recently called by Gunderson "the Investigational Asymmetries Problem" and "the Characterization Problem." I will conclude with some remarks on the bearing which evolutionary doctrine seems to have upon each of these problems.

If there is one conclusion pertaining to man about which the theory of evolution has removed all reasonable doubt, it is that men now on the earth belong, without exception, to a single species. Mankind is zoologically one, the sole surviving species of the genus *Homo*, so that we are all alike in having the same set of chromosomes required to constitute our membership in the species, and even in having the same number of chromosomes (with some rare exceptions).

The strong similarity among all representatives of homo sapiens is reinforced by the results of such life sciences as human anatomy, physiology, histology, etc. In the light of this, it is reasonable to believe that whatever method or methods of investigation I can use to find out about the characteristics of fellow-members of the species can also be used to find out about my own characteristics. For there is nothing so unique about me as to set me quite apart from them, and hence to provide me with a basis for applying a unique method of investigation to myself alone.

Yet it is precisely this last point which many philosophers have questioned in investigating mental phenomena. For many have held and do hold, despite the arguments of people like Ryle and Wittgenstein, that I have direct, privileged access to my own thoughts, feelings, imaginings, decidings, etc., and to my "subjective self," and that I can have no such access to the mental phenomena of others. Therefore, the method I use to find out about my mental states (e.g., classical introspection) is totally different from the method I use to find out about my fellows; and the same applies to each of them in relation to all the rest. From this conclusion it has been common to affirm a dualistic ontology that sets up a conceptual tension between this affirmation and the ontologically monistic implications of an evolutionary view of mind. Moreover, where the evolutionary approach allows men to conceive of themselves as within the natural world that is their home, the other approach requires, or at any rate encourages, what Whitehead once called "a false dichotomy of Nature and Man." The weight of the evolutionary approach is thus on the side of those like Gunderson who seek to show that the investigational asymmetry associated with some mental phenomena does *not* "thrust upon us any dualistic ontology of things, processes, or features."

Does evolutionary theory have any significant bearing on the so-called "characterization problem," which Gunderson describes as "the problem of providing an adequate inventory and anatomy of those features which we, in fact, find other persons to possess"? This is too large a question to consider at the end of a paper. Instead, I will simply mention one line of thought which future inquiries might well pursue.

If we consider the phylogenetic history of the nervous system in terms of concepts of information-theory and cybernetics, we can say that it has been marked by a steady increase in efficiency for collecting, storing, and transmitting information. Much of this information has been about ways to keep alive in a wide variety of habitats. Now the increase in efficiency is largely a result of the increase in complexity of organization of the *systems* concerned, together with the increase of what, in cybernetics language, are called "restraints," factors which determine inequalities of probability. Natural selection is itself one of these restraints, as are the local environments of individual populations. But this view leaves no sharp boundary between an organism and its environment: A living organism is a thermodynamically "open," steady-state system. Where it is equipped with a brain and sense-receptor organization, the system involved can be thought of as *an organism-environment whole* which functions causally, but with a reticulate, not a linear, mode of causal action.

Now, we might find that we would gain increased understanding if we envisaged the evolution of mental phenomena, or of mental phenomena and neural complexes, as a *sequence of systems* marked by steady growth of complexity in their organization. This would square with the claim I have advanced that such systems had from the outset a positive adaptive advantage to their possessors and were favoured by natural selection. In the case of our own species, if we were to think in terms of a *mental-neural network system*, having a complex array of functions, instead of in terms of mental substance and psychical events, or of physical substance and bodily behaviour, we might generate the fresh approach to issues in the philosophy of mind, which, I suggested at the start, now seems to be needed. Such an approach would bring philosophy of mind into the post-Darwinian age.

Bibliography

Darwin, Charles, *The Origin of Species,* First ed. 1859, Sixth ed. 1872.

—— *The Descent of Man,* 1871.

Gray, Asa, *Darwiniana,* 1876.

Gunderson, K., "Asymmetries and Mind-Body Perplexities," *Minn. Stud. in the Philos. of Sci.,* IV, 1970.

Haldane, J.B.S., *The Causes of Evolution,* 1932.

Huxley, J.S., "Higher and Lower Organization in Evolution," *Journ. of Royal Coll. Surg. of Edin.,* VII, 1962.

Oakley, K.P., *Man the Toolmaker,* Second ed. 1952.

Popper, K.R., *Objective Knowledge: An Evolutionary Approach,* 1972.

Rensch, B., *Evolution Above the Species Level,* Eng. tr. 1959.

——*Biophilosophy,* 1971.

Romanes, G.J., *Mental Evolution in Man,* 1888.

—— *Mental Evolution in Animals,* 1883.

Ryle, G., *The Concept of Mind,* 1949.

Thorpe, W.H., *Learning and Instinct in Animals,* 1962.

—— *Animal Nature and Human Nature,* 1974.

Wallace, A.R., *Contributions to the Theory of Natural Selection,* 1870.

Wittgenstein, L., *Philosophical Investigations,* 1953.

Scientific Objectivity and the Mind-Body Problem

JOHN T. O'MANIQUE

> I knew that I was a substance the whole essence or nature of which is to think, . . . so that this 'me,' that is to say, the soul by which I am what I am, is entirely distinct from body, and is even more easy to know than is the latter, . . . (Descartes, *Discourse on the Method,* Part IV).

With the same mathematical certainty found in this assertion that his mind is a thinking substance, Descartes will soon conclude that his body is an extended substance and will proceed to deduce a system of truths from these clear and distinct principles. All can be known with certainty. The thinking substance, the "I" itself, is lodged in the pineal gland from whence it can use its God-given mathematical aptitudes to understand fully a world created for the geometer.

Some 330 years later the clarity, distinctness, and certainty is lost, except within deductive systems which are now known to be closed to the world. Knowledge increases exponentially, but the unknown accelerates before it at a still greater rate, leaving the epistemological optimism of the seventeenth century far behind—a distant and very strange dream. The "mind" is more mysterious than ever, the nature of matter is not clear, and uncertainty has the status of a scientific principle.

Our century is, of course, a time of general uncertainty, and uncertainty in many areas may indeed be disturbing. For the seventeenth century to have solved all problems, however, might also be disturbing: Having solved a puzzle and admired it, what does one do then? We may even be thankful that Descartes was mistaken, that problems do exist, and that some of these problems, like the problem of knowledge itself, can still exercise us and lead us in new and interesting directions.

The purpose of this paper is to look at the problem of certainty and the related issues of objectivity and the nature of the mind. The first part of the paper will present an historical and rather abstract overview of Western man's quest for certainty and its attendant assumptions. The second part will focus on the radically new perspectives on these matters arising from contemporary physics.

Footnotes to Plato

Descartes is often called the father of modern philosophy; and indeed, he deserves the title, for much of his thought was original and has had a major influence on all of Western thinking to the present. Descartes, however, is also the son of ancient and medieval philosophy: His thought, as one should expect, is not a complete break with the past— in fact, in many significant respects, it is a continuation of a traditional Western world-view containing three fundamental elements: matter, including the human body; the immaterial element, the foundation for being and activity; and the link between the two—the soul or mind.

One of the first, if not *the* first, clear statements of this triad is found in Plato as the basis of his ontology and subsequently of his social theory. The transitory nature of material bodies and the permanence of certain ideas led Plato to the conclusion that there are two worlds: the sensible world of generation, change, decay and imperfection—the world of matter; and the ideal world of permanence and perfection—the world of forms. The material world is unreal, a pale shadow of the real world of forms. Its existence is no more than a participation in the Ideal.

The mind, or soul, is the link between the two worlds. It originated in the world of forms and hence, in some way and to some extent, participated directly in that world. But, as Plato tells us in the *Phaedo*, while in the world of shadows, the soul is simply fastened to the body, viewing existence through prison bars, wallowing in the mire of ignorance. The philosopher, nevertheless, can reactivate some of the original participation, can bring himself and his students closer to the Ideal, even before death finally releases the shackled mind and permits its return to the world of forms.

Aristotle accepted the Platonic position that clear and certain knowledge required an object that is real and unchanging, that is, true knowledge is knowledge of the forms. Aristotle, however, integrated the world of forms with the world of matter. Prime matter, the principle of change, potency, individuation, and substantial form, the principle of act and perfection, are essential principles of all finite being. Left in the "other world" is the Prime Mover, the pure act, the source of all perfection.

For Aristotle, the mind or soul is the form of the human being, the principle of its vegetative and animal life, and of its rationality. As principle of rationality or knowledge, the human form is related to other forms as the senses are to sensible objects. It is not simply a passive receptor of forms, however, but, through its form, actively knows the object, judges, and reasons. The activity of the rational form is in some way, not clearly defined by Aristotle, a participation in the activity of the Prime Mover.

This very brief account of the philosophies of Plato and Aristotle is presented to introduce the traditional view referred to earlier. The common elements in the epistemological theories of these two founders of Western thought are clear, objects of knowledge commensurate with the human mind. Man is, therefore, ultimately capable of clear, distinct, and certain knowledge. This commensurability, and hence the certain knowledge that flows from it, is divinely guaranteed. The guarantee is fulfilled through some kind of participation of the human mind with the divine. The mind and the intelligible aspects of its objects, the forms, share, therefore, in the immateriality of the divine; the sensible world, the realm of uncertainty and change, is the world of matter.

Throughout the various philosophical developments of the Middle Ages, emphasis shifted between Platonic and Aristotelian versions of this basic position, but the position itself survived. In fact, it not only survived, but was given greater strength through the support of the personal God of Christianity. God creates the objects to be known, makes the soul that knows in His own image and likeness, and guarantees the ultimate possibility of truth and certainty.

Descartes started from this position, but made major modifications in it. The mind, the Cartesian thinking substance, is immaterial, and even more distinct from matter in that it is itself a substance, not simply a principle within a substance as in the Aristotelian tradition. The existence of God is deduced, as is the proposition that God guarantees certainty. The nature of the guarantee, however, is different: The mind does not know through participation in the divinity and a resulting conaturality with the forms, but because of the nature of the mind and of the material world itself. Matter is extended substance, the object of mathematics, and as such fulfills the Platonic requirements for intelligibility. The mind is the mind of the mathematician. Whether God created them this way or not (and Descartes apparently believed He did), an active and continuing participation in the divine is not required for certainty. Divinity fades away as the immediate foundation for knowledge. Matter is no longer a shadow unintelligible in itself. It is, rather, like any definition, axiom, or proposition of a mathematical system eminently intelligible, and its intelligibility is exhaustible. Certainty is not merely an ultimate possibility but, given the correct method—the Cartesian method—it can be a reality here and now. And the foundation for this certainty is not God, but the conaturality between the thinking substance and the rational universe.

Epistemological optimism had reached its zenith. Other thinkers of the period—Bacon, Galileo, Hobbes—shared Descartes' vision of a world written in the language of mathematics, just waiting to be clearly and distinctly understood through the mathematical method. But the optimism was to be short-lived. Rather than solving all of mankind's problems, Descartes created some of his own, one of the most significant being a very difficult version of the mind-body problem. As long as the mind and its objects were believed to participate in the divine, the exact workings of the mind and its relationship with its own body and material bodies in general would not be seen as a critical problem. In such enquiries where reason failed, faith took over. Thomas Aquinas, for example, saw angels as agents informing the mind. The Cartesian deus ex machina, however, appears between the proofs for the thinking and extended substances, then disappears to leave the rest to the unaided reason. And that unaided reason could not adequately explain the interactions between the substance, mind, and the clearly distinct other substance, body. Attempts at solution involved elimination of one or the other of the two substances, that is, led to either material or spiritual monism. Much of modern Western philosophy is a continuation of this Cartesian bifurcation and, of course, cannot all be dealt with here. Because of its foundations in early physics, and its continued great influence to the present, the development of material monism alone will be looked at in this paper.

The Realm of Point Masses

The scientific principle that seems to have given extended substance the ultimate edge over the thinking substance was the new principle of inertia, as introduced by Galileo, and established as the first law of motion by Newton. Before Galileo the common sense notion of inertia was the only notion of inertia: Inertia was the property of material bodies whereby they resisted motion. Matter was seen to be inert, inactive in itself, hence the need for immaterial principles and teleological laws. The new concept of inertia was a radical departure from the old, and also, therefore, from common sense. Inertia became the property of a material body which resisted a *change* in motion (an acceleration). Rest was no longer the natural state of material bodies; constant velocity was.

Many saw the material world as nothing more than point masses moving in space—point masses which could be mapped isomorphically onto mathematical systems, which had a new-found inner dynamism. This Cartesian-Newtonian world view—that of Classical Mechanics— left little room for the thinking substance. Could not the workings of the human mind itself, like everything else, be explained in terms of point masses moving in space? The explanation would surely be complex, since the mechanical system being explained was complex, but that would simply be a matter of degree.

Knowledge was no longer rooted in the divine; theism gave way to

110

deism and then, through the method of mathematical physics, the world, including all of man, became eminently knowable without any immaterial intervention. Auguste Comte applied the method to the social sciences, John Watson to psychology. The theory of evolution gave added support to the view of a homogeneous, continuous nature, and hence to the Cartesian-Newtonian model as the explainer of all nature. Throughout, despite the major changes, one element of the traditional view survived. The dualism that claimed the mind to be immaterial, the body material, was generally dying, and was, in positivistic-scientific circles, dead. Nevertheless, another fundamental dualism lived on—the subject-object dualism.

The human mind, in spite of the loss of its divine character, was still in a privileged position, the privileged position of the classical physicist. The knowing subject could know the objective world in an objective way. In other words, the world was still intelligible and could be known *as it is*. The subject was still separate from the object.

The positivistic world-view was not, however, without its obstructions. The optimism of the seventeenth and eighteenth centuries, some of which has continued to the present, was from the outset confronted by serious questions about the ways in which man knows and the validity of his knowledge.

If knowledge, for example, is no more than a complex mechanistic process with no divine guarantee, if it is no more than the reception by the mechanistic "mind" of point masses collected through the senses, then what can be said to be truly known except these point masses—these atoms of evidence or sense perceptions *within* the knowing subject? The answer found within the logic of Locke's psychology is that nothing else can be known, though Locke himself most inconsistently avoided the consequences of his own logic. (He spoke, for example, of the primary, objective qualities of bodies as if they were objectively known.) Berkeley accepted the logic; to be *is* to be perceived. But this principle of extreme subjectivism was used by the Bishop to confirm his belief in God, the divine guarantor of knowledge. In such a context, who need worry about the mechanics of the knowing process? If God is the ultimate source of both the pad of paper before Bishop Berkeley and his mind which contains the perception of it, he can be as secure in his belief that paper is really there as Thomas Aquinas would be with his somewhat different version of divine intervention.

David Hume, the most logically consistent of the great British empiricists, accepted the inevitable subjectivism without the divine guarantee. The sense data were all that could be known. Hume did not, however, fall into the depths of skepticism. He *believed* in certain

111

things that cannot be *known*; for example, he could not *know* causality for there are no sense data of causality, but only of successions of events. He, nevertheless, believed in the principle of causation, the foundation for the belief being, somewhat anachronistically, a principle of natural selection. And the belief may be, he stated, more infallible than our knowledge.

In response to Hume's skepticism, Immanuel Kant presented for our consideration a new and brilliant model of a mind that was designed with categories of an absolute character to know things as they are. "Things" here refers to phenomena, "things as known," not noumena, "things in themselves." This ruled out metaphysics, but, since the Kantian phenomena have objectivity, it also ruled out the subjectivism of the British empiricists. Kant rejects Descartes' extreme dualism, arguing that the mind can be known only against the background of sense experience, nevertheless he does retain a moderate body-soul dualism, as well as a belief in God which, though not rationally proved, can be rationally justified. The originality of his view notwithstanding, Kant preserves the possibility of true knowledge of the real world by returning to the spirit of the Platonic-Aristotelian tradition.' It would appear, at this point at least, that some form of dualism that posits a mind designed to know the material world around it is a prerequisite for objective knowledge, and that some kind of supernatural guarantee is implied. A purely mechanistic model, without any type of ghost in the machine, leads to subjectivism and skepticism.

This return to objective realism, however, was not to go unchallenged. The nineteenth century introduced the most devastating attacks on the foundations of objective knowledge. A paradigm case, Friedrich Nietzsche, pronounced the divine foundation of knowledge dead, and went on to destroy any human basis for objective truth by equating scientific systems to poetry and myth. Both are metaphorical systems; the only difference he saw between them is that man believes the sets of scientific metaphors to be true. For Nietzsche the divine foundations *and* the rational foundations were gone; the absurd remained—the confrontation between the rational animal and the irrational universe (to use Camus' definition of the absurd).

Many continued to believe, in spite of the Humes and Nietzsches, that there is such a thing as objective knowledge. If so, what guarantee is there of certainty?

The early seventeenth-century optimists believed that a good mathematical model mapped onto the real world and through the mathematical method all could be known with certainty; however, the extension of this method itself proved otherwise. The nineteenth-century discovery of non-Euclidean geometries demonstrated that the logical perfection of a mathematical system is no guarantee that it can be mapped isomorphically onto the real world. Within Euclidean, Lobatschewskian, and Riemannian geometries, it is *true* that the sum of the angles in a triangle is equal to two right angles, less than two right angles, and greater than two right angles, respectively. Can all three be true in the world *outside* deductive systems?

This question, and, of course, its answer, were critical within the development of science itself. A perfectly logical set of ideas was no guarantee of objective truth; deductive systems need not tell us anything about the real world. In fairness, it should be noted that Descartes himself said that, when dealing with complex issues in physics, an empirical test may be required to determine which mathematical model actually "mapped onto" a particular set of events, and also that the Kantian a priori forms need not be so specific as to rule out different geometries. There was, nevertheless, the general belief, not shattered until the nineteenth century, that a good mathematical system must map onto the real world, and hence, mathematical truth was truth about the real world.

This discovery, coupled with the growing recognition that in any application of mathematics to the real world certainty is lost, destroyed any hope of fulfilling the Cartesian dream of absolute truth. Nevertheless, within the scientific community there still remained a faith that, if not in practice, at least in principle objective and certain knowledge was possible.

Waves and Uncertainty

During the early years of this century there emerged from physics a radically new world-view. The science giving rise to this different perspective was, however, much more esoteric than the physical foundations of the mechanistic world-view, and furthermore, it flew in the face of common sense much more than did the principles of Galileo and Newton. Consequently, the new physics, relativity, and quantum mechanics, have not had the impact or general influence on other areas of knowledge that one might have expected from so fundamental a shift in perspective. Without an attempt to explain these physical foundations in any detail, we will look at some of the bases in relativity physics and in quantum mechanics of this new and strange picture of man-in-the-world.

As suggested above, the Newtonian definition of inertia as the property of a body whereby it continues to move with constant velocity contradicted the common sense notion that matter, by itself, had no dynamism at all—was inert in the usual sense of the word. Einstein's famous energy-mass equivalence equation ($E = mc^2$) conflicts with common sense much more than did Newtonian physics. It states that mass—the quantitative measure of inertia—is equivalent to energy—the capacity for work or action—and that the energy of a body equals the mass multiplied by the square of the velocity of light, a truly enormous quantity. Everyone is familiar with a most unfortunate application of this discovery, the nuclear bomb, and a more fortunate but still questionable application, the nuclear power plant, is gaining fame. It is not generally recognized, however, that every bit and piece of matter around us has, or more precisely, is a large quantity of energy, easily calculated by use of the equation. The only difference between the material in the atomic reactor and in the pencil lying on the desk is that the energy of the latter is much more difficult to release, a difference for which we might be thankful. Matter is bound-up energy, and energy is the capacity for action. There is less need than ever before for an immaterial principle to explain activity.

Early twentieth-century images of bound-up energy were essentially in the classical mechanics' mould. The Rutherford-Bohr model of the atom presents the familiar picture of the large inner sphere, the nucleus, surrounded by smaller spheres, the electrons, revolving around the nucleus in orbits as the planets revolve around the sun. The nucleus itself is seen to be tightly-packed spheres, neutrons, protons . . . , held together by the huge nuclear forces which account for the tremendous quantities of new-found energy that is in every bit of matter. This new-found energy, however, could be simply added to the point-masses-moving-in-space of classical mechanics. The world is still pictured to be, at base, tiny particles in various configurations, but now having (or being?) larger quantities of energy than was previously conceived.

The breakdown of the purely corpuscular or particle picture of matter is a development of quantum mechanics. It began when Max Planck hypothesized that all energy exchanges involve discrete unit packets of energy, the energy quanta (just as a piano string will vibrate with only discrete frequencies and hence will have and exchange only the discrete energies proportional to those frequencies). Einstein identified the photon as the fundamental quantum of energy. But what are these packets of energy called photons, electrons, protons, etc.? Are they tiny particles which "have" great quantities of energy? Are they particles with discrete frequencies analogous to the wave motion of the piano string? Are they not particles at all but simply wave motions . . . ? Further experimental work in quantum physics provided an answer to these questions: "It depends . . . ."

114

A beam of "packets," say photons, travelling from a source through a small slit in a screen and impinging on a second screen, will produce an effect that can be explained by assuming that the photons are tiny particles. The introduction of a second slit close to the original one in the first screen, however, produces an interference pattern on the second screen which can be explained only by assuming that photons are waves. This is not an isolated instance: In general one set of phenomena associated with the fundamental energies could be explained within the context of a particle model, another set by a wave model. Bohr developed a sophisticated mathematics to bridge the particle and wave models and, in layman's terms, stated that the fundamental energies could be considered to be either particles or waves, depending on the experiment being performed (hence the "it depends" mentioned above). This ambiguity and incongruity was too much for some scientists still used to clarity and distinctiveness, and a few opted for either the particle or the wave model as the *true* representative of the "real" world.

Heisenberg's Uncertainty Principle (or Indeterminacy Principle) shed further light, or darkness, on the problem. Abstracting again from its complex mathematical and experimental foundations, the Principle states that in any measurement of fundamental energy, the product of the error in the measurement of the position and the error in the measurement of the momentum (mass times velocity) will always be equal to or greater than Planck's constant (h), that is, $\Delta x \cdot \Delta mv \geq h$, where h is the constant in the formula for the fundamental quantum of energy—the energy of the photon (h times the frequency of the photon equals the energy of the photon).

This would appear at first sight to be a purely epistemological principle; it seems to deal not with the "real" world but with errors in measurement. Heisenberg himself and those subscribers to the Copenhagen interpretation of the principle, argue that this is an ontological principle; the Uncertainty Principle tells how things really are, but at the same time changes the meaning of "how things really are."

There would appear to be two reasons for the Copenhagen interpretation, and because of their importance in dealing with the major issues in this paper, they will be looked at in some detail.

First, uncertainty arises because any measurement involves an interaction between the subject measuring and the object measured, and the interaction changes the object. The changes in most cases of measurement at the macroscopic level are negligible. At the atomic level they are not negligible, and this is not simply a matter of the present state of the measuring techniques. For some time it had been recognized that, in practice, because of the limitations of the scientist and his instruments, there would be errors in measurement. But, it was also believed that, in principle, these errors could be eliminated—zero error could be approached indefinitively through the refinement of tools and techniques; in other words—the traditional belief that, ultimately, certainty could be achieved. Heisenberg said *no*; even in principle, the limits of accuracy set out by the Uncertainty Principle cannot be exceeded. An analysis of measurement itself may make his position clear.

Any measurement, indeed any observation, requires a transfer of energy, an interaction of energy fields. Seeing a table, for example, requires first an interaction between its surface energy (in the form of its atoms) and another energy, in this case light (photons). Both the light-energy and the table-surface-energy are changed somewhat by this interaction. Certain frequencies of light-energy, for example, will be absorbed, changing the energy levels of the surface-energy. Some light-energy will be reflected by the surface and will be received by the observer's eyes. The changes in surface-energy-levels produced by the impinging light will have some (probably negligible) determining effect on what light is reflected (not negligible if one observes a piece of paper with a magnifying glass in bright sunlight and sees it burst into flame as a result of the observation). The "object observed," then, is not simply the surface of the table, but the surface-light interaction. In order for one to see anything, an interaction between the "object's" energy and light-energy is required, and the light is a part of the subjective side of the process—in fact, the kind of illumination used may be determined by the knowing subject. More will be said later about the subjective side of an observation.

If the examination of the table involved feeling it, the required interaction, which is what would be observed, would be between the energy of the table-surface and of the surface of the fingers, and again, some change in each would result.

Now let us consider an observation at the micro-level—the observation of a sub-atomic particle, say an electron. Energy is required, as in the case above. If one wanted to "see" an electron in orbit around a nucleus, just as in seeing a table one would have to bombard the electron with some form of energy (or, to avoid the particle language, produce an interaction between the quantum of energy called electron and some other energy). The detail that one sees depends on the length of the energy wave used by the seeing subject relative to the size of the object seen. Radar, for example, using long radio waves, will not enable one to see details smaller than a few feet in extension. The shorter the wave-length of the energy used, the finer the detail observed. For the observation of macroscopic objects, all wave

116

lengths of visible light are very small relative to the size of the object observed. Relative to the size of the electron, however, they are not small, but of the same order of magnitude.

Nevertheless, short wave length gamma rays (not visible light) provide sufficient resolution to allow some observation of the electron, (roughly comparable to observing an aircraft on a radar screen). To make the observation, however, at least one gamma ray photon must be reflected from the electron, and in the process the electron is knocked from its orbit. Data about this interaction could be collected and would tell us something about electrons, but what is known about electrons would have been abstracted from what is known about the electron-gamma ray collision. It is impossible, as this example shows, to "see" the electron in orbit, or in any motion, for the "seeing" itself alters the motion, and hence, its momentum.

The above deals with what might be called the objective side of the observation. The question still remains: How does one collect the data from the interaction itself? A procedure must be designed whereby certain changes due to the interaction and in the properties of the energy chosen by the subject (in this case, the gamma ray) can be determined (position, momentum, trajectory, energy level, etc.), and from these, with the use of theoretical models, something can be said by inference about the electron. This discussion of the "subjective" side of the problem will be continued later.

The second major reason for the belief that uncertainty and indeterminacy are real properties of matter and that they therefore cannot be overcome in principle, is the wave nature of matter. Cartesian certainty, at least in the world of extended substance, depended upon the reducibility of matter to point masses whose positions could be plotted accurately on the Cartesian reference frame, and as has been seen above, this was the project of Classical Mechanics. If the basic stuff is not particles but waves, then position in the Cartesian sense becomes meaningless. As Heisenberg describes it, a "wave packet" is a wavelike disturbance whose amplitude is appreciably different from zero only in a bounded region which is itself in motion—and which changes its shape and size. This somewhat arbitrarily bounded and changing region could not be identified with *a position*, if position refers to a point in space. Position could be used only to refer to an area defined by the Uncertainty Principle, within which the "wave packet" would probably be found. Any use of the word "position" (or "velocity") with an accuracy exceeding that of the Uncertainty Principle is, according to Heisenberg, as meaningless as the use of words whose sense is not defined (Heisenberg, 1930).

The data received by the knowing subject are results of interac-

117

tions between the energy of the object known and the energy used by the subject to observe; these interactions change the object; the precise identification of position is a meaningless concept. These are the reasons presented by the Copenhagen interpretation for rejecting the possibility of knowing things *as they are* and the possibility of knowing anything with certainty. This rules out not only the Cartesian-Newtonian world-view, but also the essential epistemological elements of the traditional Western world-view, and of common sense. The concept of things *as they are* itself becomes meaningless. A thing *as it is* is unknown; a thing known is a thing interacting with a knower. (It should be noted that those who may still hold out against the Copenhagen interpretation that all of this is the case *in principle*, would probably agree that it is the case in practice, and hence, at best, would see great problems in attaining certain, objective knowledge.)

As suggested more than once above, our common sense world-view is similar to traditional Western philosophical world-views, and to that of a Classical physicist. As a result, our concepts and the way we express them in language are, for the most part, compatible with the world-view of Classical mechanics, especially if some kind of ghost is added to the machine. Western man thinks in terms of things (substances) doing, having, being . . . , in other words in terms of subjects and predicates, substantives and verbs. These related dualisms are more fundamental, it seems, than the mind-body or the subject-object dualism, but all are probably interrelated. Because the very language that we must use to discuss the new world-view is itself based upon the old world-view, the task is very difficult (if not impossible). Nevertheless, some attempt will be made to develop the implications for the quest for certainty of this new view further, with reference in particular to the dualisms just mentioned.

On Falling Pencils

The traditional and common sense world-view sees material *things* (rocks, oranges, dogs, humans . . .) as more or less solid, stable, invariant substances composed of parts which are themselves stable, invariant substances. Changes were and are always noted, but the emphasis is on stability or invariance.

Some of these *things*, including human beings, could know other *things*, and human beings could know others in a peculiarly human way. To summarize what has already been stated, this knowing required, according to many philosophers, an immaterial principle or part within the human being, but whether or not this was the case, knowledge of the object known usually involved the reception of data from that stable, invariant object out there, data which, when processed by the knowing subject, represented the object *as it is*. The knowing process did not change the object known, and the data from the object were adequate to the job—they could be a basis for an exhaustive knowledge of the object.

This latter belief was usually guaranteed by the very design of the knowing subjects, and further by the designer. In the Platonic-Aristotelian tradition, the designer-guarantor was a divinity, the prime-mover and eventually the God of Christianity. From the seventeenth century on, it was the rational-mathematical nature of the universe, as designed by the God of Christianity, or a less personal god, or simply a mathematical universe itself. The importance of this type of belief in ultimate design was underlined by Einstein when, in rejecting the Copenhagen interpretation of the Uncertainty Principle, he stated that "God may be sophisticated, but He is not malicious" and "I cannot believe that God plays with dice." These statements need not imply any particular belief in God on the part of Einstein, but they do show his belief that the human mind is commensurate with the design of the universe.

In sum, according to the traditional view, the subject and object are distinct, invariant substances, the subject receives data from the object out there without changing the object, and there is a guarantee of the veracity of those data.

The contemporary world-view that arises from new concepts of matter-energy equivalence, from the Complementarity Principle with emphasis on the wave model, and from the Copenhagen interpretation of the Uncertainty Principle, rejects each of these elements of the traditional world-view, as we have seen above. In doing so the basic concept of *things* itself is radically modified. What appears to the knower to be an invariant substance made up of invariant parts is, in this new model, a part of an interaction between two sets of indeterminate wave activities. This statement must be explained in some detail, but first it should be noted again that the concepts and language used to explain the new model are the concepts and language of the old model, which the new model rejects—hence some of the difficulties in explanation.

To make this analysis more intelligible, a simple event will be used as an example: a falling pencil. We first take the position that to deny that one could know that a pencil is falling would be foolish. The human species has survived using the traditional and common sense views and, within the context of natural selection, it would seem that these views of knowledge that are being questioned have great survival and hence practical value. If Plato, Descartes, Hume, or Heisenberg need a pencil, each will reach out and pick up what each *knows* to be a pencil in essentially the same way. Every normal human could know that a pencil is falling. We need not question that; we wish, however, to analyse the meaning of "knowing that a pencil is falling" in light of the wave model as presented above.

Consider first the proposition "The pencil is falling." This statement is in the usual subject-predicate form, the subject referring to what is perceived to be an invariant substance or *thing*, the pencil, and the predicate to an activity or behaviour of this *thing*, that is, falling. The subject-predicate dualism in this case is also a substance-action dualism or, a *thing*-behaviour dualism. Does this dualism stand up to analysis within the new model?

What is known of the pencil in terms of the new model could be analysed considering what is seen by the unaided eye: colour, shape, texture. . . . The surface-energy of the pencil is interacting with light-energy (as outlined above with reference to the table). These interactions or energy transformations depend upon the energy levels or frequencies of the waves of the surface-energy and of the light-energy. Some of the light-energy will be "absorbed" by the pencil, some reflected according to the quanta of energy involved. The light-energy absorbed will, of course, change the surface, and this change will determine what light-energy is reflected and what other energy is radiated from that surface. (Wherever the word "determine" is used in this context it must be taken in the soft sense, compatible with the Uncertainty Principle.) This would become obvious to the observer if strong sunlight is focused on the pencil by a magnifying glass, or more so if a laser beam is used for illumination (the latter affecting more than the surface-energy). Under unfocused, cool light changes like this would not be detected by the naked eye; nevertheless, the light reflected would be conditional upon the energy levels of the surface as altered by the impinging light.

According to the Uncertainty Principle, no precise determination of these transformations is possible, and, in the Copenhagen interpretation, the wave packets themselves do not behave in a determined way. One could not predict with certainty that a particular wave packet with energy $c_1 f_1$ in the surface of the pencil will be at a particular point x_1 at time t_1 and will interact with a particular light wave-packet with energy $c_2 f_2$ to produce a particular result. Given, however, the very large numbers of wave packets involved, the probabilities of large numbers come into play; one can predict, if not with absolute mathematical certainty, nevertheless with a sufficiently high degree of probability for all practical purposes, that a large number of transformations of a particular type at the micro-level will occur to give, for example, the appearance 'yellow' at the macro-level. There may be a statistical possibility that the yellow pencil could disappear from sight, or turn blue, but this probability is so close to zero that one can be practically certain that a yellow pencil will remain yellow. (By "practically" we do not mean "almost" but rather "for all practical purposes" as opposed to "mathematically certain.")

To move to the subjective or knowing side of this analysis of seeing the pencil: Some of the reflected and radiated energy will impinge upon the eye of the observer. The energy-levels of the optical system (pupil-lens-retina-optic nerves-cerebrum) will provide the conditions for what part of this total energy from the pencil is received and processed. Only those frequencies within the narrow band of the electromagnetic spectrum known as visible light will interact with the optical system; other frequencies will not be received by this system and hence will not be a part of "seeing," although they may be even consciously received (e.g. the reception of radiation which is sensed as heat).

At the subjective end of the process, just as at the objective end, the interactions produce changes in both the received energy and in the receiving energy, that is, in the energy that makes up the appropriate parts of the nervous system. Seeing is, it would seem, nothing more than the resultant of some of these changes.

The same kind of analysis could be applied to other types of observation. Tactual observations, for example, involve the interaction of energy waves in the surface of the pencil with those of the observer's fingers, then through the nervous system to the appropriate parts of the brain. As above, both subject and object are modified by the interaction, and the part of the total interactions that is "known" is determined by the knowing apparatus.

In any such analysis, *things* do not emerge; that is, one does not find a solid, invariant entity which corresponds to our traditional or intuitive notion of *thing*—or substance. All that is discovered is behaviour or activity. Waves are activities: There is no need to postulate a medium (the ether) which *has* the wave motion. The closest thing to a *thing* (this is an example of the language problem) is the wave packet which is, in the sense identified earlier in this paper, locatable—but it may appear to be nevertheless, a pure wave motion.

The statement that "there are no things" seems, at first sight at least, ludicrous because it is so alien to common sense. We might recall, however, that earlier, less radical discoveries which conflicted with common sense were also considered ludicrous at the time—for example, the discovery of the Galilean-Newtonian principle of inertia. What is being stated here must be made as clear as possible: "There are no things" means that within the context of this new world-view, there is no reason to posit the existence of entities that are invariant, solid bits of stuff which traditional and common sense concepts of *thing* map onto. To say "If there is activity, some *thing* must be acting" is a statement based on that traditional and common sense way of seeing the world. This way of conceptualizing the world is good and practical—it works; but that in itself is no guarantee that this conceptualization maps perfectly onto the ultimate whatever-it-is out there. This new view suggests, in fact, that there is no guarantee that any set of concepts will map perfectly onto the extramental world, and that such a notion may be meaningless.

What is known as a *thing* at the macro-level is a relatively invariant set of wave-packets, a set of activities producing, because of their large numbers, a relatively stable statistical result. The pencil lying on the desk is seen to be a stable, invariant thing. If it is left lying on the desk at night, one would expect to find it in the same place, with the same shape, colour etc. the next morning. It could have moved. Its molecules are in constant motion, and it is possible that a net movement to one side would be sufficient to overcome the friction between the pencil and the table, or a net movement up to overcome the force of gravity. In other words, it seems that the movements of wave packets and the more complex structures they make up are not so determined that their resultant kinetic energy is always zero; nevertheless, given the large number of movements, the statistical result will always approach zero. If one thousand dice are thrown, one thousand two *could* come up, but the odds in favour of that are $1:6^{1000}$, so slight that we are justified in calling the possibility an impossibility. It could happen, but what we expect to happen is that approximately 166 twos will come up, and that approximately 166 of each of the other five numbers will also come up, producing a fairly well balanced result.

The point, then, is *not* that the view of variance and indeterminacy at the micro-level should cause us to abandon our view of invariance and determinancy at the macro-level. The common sense view works quite well for all practical purposes. The point *is* that we should review the philosophical consequences of the new view versus the common sense and classical mechanics views, and particularly the position that knowledge is not the reception of data from stable, invariant *things*, but rather a part of a complex set of interactions among variant activities.

Let us consider the predicate of the sentence "The pencil is falling." Within the common sense view, the only activity referred to is the falling; the pencil itself is the invariant *thing*. In light of the above analysis, there would seem to be no essential difference between the noun "pencil" and the verb "is falling" in terms of a *thing-action* dichotomy. The falling is an activity resulting from an interaction between the pencil and the Earth. But the pencil itself is no more than a complex set of activities. The verb refers to an activity that is added to a relatively stable set of activities, referred to by the noun. This would seem to be the general case for nouns and verbs. Subjects refer to relatively invariant sets of activities and predicates in general refer to some set of activities abstracted from or added to the invariant set. A property would be a sub-set of activities abstracted from the whole set. For example, in the proposition "The pencil is yellow," the predicate refers to the sub-set of interactions between the surface of the pencil and light which produces the sensation yellow. Subjects *and* predicates, whether the latter are verbs or adjectives, are all "action-words," that is, all refer to activities or behaviours, and to nothing more—at least within the context of the wave model.

This analysis of a somewhat trivial set of behaviours, the falling pencil, can now be applied to a more problematic set of behaviours—those related to knowledge itself. Let us consider the proposition "He knows that the pencil is falling."

The subject "he" refers to a relatively stable complex set of activities. The verb "knows" (or "thinks," "is aware," "is conscious," "believes" . . .) is an activity which is possibly related to the whole set of activities referred to as "he," but probably is concentrated in a sub-set of those activities known as the nervous system, and, in particular, in the cortex of the brain. If, as in the traditional view, the brain is a *thing* or a complex collection of *things* (for example, neurons as point masses moving in space), then the traditional mind-body problem emerges. Early concepts of material things would lead to the conclusion that the material brain itself would not be adequate to the task of performing the activity of knowing (thinking, being aware . . .), and hence an immaterial principle would also be required. A concept of a more dynamic matter, as is the post-Galilean concept, may lead some to the conclusion that a separate thinking substance or principle is not required, but the relationship between the act of knowing and the thing that is knowing is still couched in a dualistic thing—action (subject-predicate) model which is difficult, if not impossible, to specify. Problems arise such as the question of the relationship between a "thought" (a mental event in general) and neurons or sets of neurons; is a thought related to one neuron, a small set of neurons, or is it a more holistic phenomenon, possibly related for instance to the cortex of one hemisphere? The problem of consciousness may not disappear within the new model, but the extra problems of the thing—action dualism—will be eliminated. A more detailed examination of the knowing process will help to clarify this.

The interactions between the pencil and the light, through the

123

resultant interactions in the optic system, produce a set of interactions within the cortex. This latter set of interactions *is* the *knowing*, just as the interaction of gravitational fields between the pencil and the earth *is* the *falling* of the pencil. The conscious experience of any *thing* is, in other words, the complex set of energy transformations—interactions within the nervous system of the one experiencing, which are the result of a chain of interactions from the *thing* experienced. It is one part of the total subject-object interaction within the subject. Just as falling is an activity added to a relatively stable set of activities which is interacting with another gravitational energy field, so knowing is a (much more complex) activity added to a relatively stable set of activities (the nervous system) which is interacting with other energy fields. (We cannot in this paper do justice to the activity of knowing because of its great complexity. We might note, however, that part of the complexity of human knowledge lies in the interacting "chains" of activity. For example, if a human is falling, the activity of falling may lead to a second activity of the experience of falling, which may lead to a third level activity, awareness of the experience of falling, that is, self-conscious activity—knowing that one knows, and, if there is time before the landing, even further "metaphysical" reflections on the whole process.)

A second knower could monitor the interactions within the first knower's cortex; in other words, the energy transformations within the cortex could be used to produce further energy transformations in a measuring apparatus and eventually within the cortex of the second knower, all, of course, subject to the Principle of Uncertainty. If the measuring apparatus is sufficiently sophisticated and the second knower is a sufficiently experienced neurophysiologist, he could know that the first knower was knowing that a pencil was falling, simply by an analysis of the data from the latter's nervous system.

This second knower's knowledge of the neural events is called, by some contemporary psychologists and philosophers, the *"objective* explanation," while the knowing associated with the neural events in the first knower is called the *"subjective* mental phenomenon." They would say that the neural events as known by the distal observer are "objective" because they are looked upon as *things* which can be measured, whereas the knowing or consciousness itself cannot be known by a distal observer—is not objective—is known only proximally by the thinking subject, hence is subjective. This subject-object dualism leads to further problems of the relationship between the two: Is the "subjective" conscious phenomenon more than the objective neural event; does the latter cause the former; can the former act back upon the latter . . . ?

Within the new model, however, the "objective neural events" are no more than the results of a further set of interactions resulting from the "subjective mental phenomena," which is the activity called knowing or consciousness itself. The "objective neural events" are a partial spin-off of the energy transformations that *are* the knowing in the cortex—in "he." Some of these transformations can, given the proper apparatus, produce transformations in another cortex of the distal observer—and through a set of translations of data and inferences, the distal observer could identify his knowledge of the neural events in the first knower with what the first knower is knowing, that is, in the given example, could identify the neural events with knowing a falling pencil. All that is involved here is a further chain of energy interactions, equally limited in that only a sub-set of the total energy is being monitored, and having the same subject-object-interaction character as had the first knowing. The difference is not the one is objective, the other subjective, but the one is distal, the other proximal.

Knowing, then, is in this context an activity added to the relatively stable set of activities known as the cortex (or the broader sets of the nervous system, or whole organism), just as falling is an activity added to the set of activities known as the pencil. And, just as the pencil's falling can be distinguished from the knowledge of the pencil's falling, so too the person's knowing or thinking can be distinguished from a second observer's knowledge of the person's knowing. The distinction does not involve a distinction between things or substances on the one hand and behaviours or actions on the other; all are activities—energy transformations and interactions. Nor does it involve a subject-object distinction; the neurophysiologists' knowledge of the neural events is just as much a subject-object interaction as is the knowing of the falling pencil. The distinction is based on the location of the activity in question (falling, knowing) within a relatively stable set of activities (the pencil, the first knower, the second knower . . .). There seems no reason to say that the second observer, the neurophysiologist, has the ideal vantage point and hence the objective knowledge. Science itself is, as Heisenberg has put it, the study of subject-object interactions; as knowledge, it is a set of subject-object interactions itself.

Uncertain Postscript

We have seen above that the set of interactions that make up our knowledge—even our scientific knowledge—is a very limited sub-set of the totality of interactions, and even this small sub-set cannot be known "as it is." The old assumption that the world and the mind that knows it were designed for one another, and hence the latter can exhaust the intelligibility of the former, may have to be abandoned. The "mind," that is, the relatively stable sets of interactions making up the parts of the nervous system most involved in the activity called thinking, through a long process of natural selection, has developed a capacity for interactions with the world sufficient for survival—a truly remarkable capacity. But it is also a truly limited capacity, and it could be otherwise. Our optical system receives, for example, a very small band of the electromagnetic spectrum—being "blind" to the rest—and could conceivably have developed to receive some other band and hence to see quite a different world.

All of this need not lead to skepticism; as stated above, we can be sure for all practical purposes that we know things and that our knowledge can lead to appropriate and fruitful action. There is, literally, more to the world than meets the eyes, but the eye-world interactions are useful and trustworthy.

Nor does all this preclude the supernatural in itself. It does reject as unfounded the traditional view of a supernatural guarantee of truth. The broader scientific world-view has for some time seen natural selection as the major immediate guarantee. On the question of an ultimate guarantor, it remains silent.

It certainly should not lead to the somewhat popular position that the world is absurd—the rational animal in the irrational universe. Some rational models of the world, some ways of seeing the world, have been questioned and even to a great extent have been rejected. But the undeterminacy and uncertainty of the new view is not irrationalism. Whatever the models used by the philosophers, scientists or by the man-on-the-street, the world goes on in its own ultimately mysterious way, unconcerned by our frail attempts to understand it. It is not the universe but man who quests for certainty.

Marxism, Social Science, and Objectivity

FRANK CUNNINGHAM

That knowledge is power has been a commonplace since ancient times. But knowledge is come by in several ways, some more effectively than others. The most effective way was systematically developed only recently in human history, at no small cost in time, energy, and even personal danger to its strongest advocates. That way is *science*, and as a dominant historical force it was primarily initiated by the bourgeoisie during and after the bourgeois revolutions in Europe in the seventeenth and eighteenth centuries.

It was then that scientists like Kepler, Galileo, and Newton were developing theories which could be tied to rigorous experimentation and observational procedures with remarkable success. The new bourgeoisie, the industrialists, and the big merchants welcomed the new science. Societies for the promotion of scientific research were set up, and prizes which today would be seen as small fortunes were offered by governments for scientific achievements. And it is no wonder: Everything the scientists touched seemed to turn into gold— or rather into profits. Work in the new physics quickly found its way into sophisticated and greatly improved navigational, mining, and manufacturing techniques and into new weaponry.

Nor were the social sciences ignored by the enthusiastic rush of science. Although condemned as atheistic and immoral, theorists such as the English and French materialists attempted to do for human society what Galileo and Newton were doing for nature, with the result that psychology and economic theories like those of Adam Smith were developed. Despite the attempts of some—for example, the English materialist Hobbes—to use their science in the service of the dying feudal order, it was clear before the end of the eighteenth century that the new science belonged to to the bourgeoisie.

How, then, in the next two centuries did the bourgeoisie react to scientific socialism? At first, and still in some quarters, bourgeois theorists attempted to meet it scientifically, by claiming that the basic views of Marx and Engels can be shown by scientific inquiry to be objectively false. But in the face of scientific thought leading to conclusions unacceptable to bourgeois theorists, especially in the social sciences, more and more of them tended to take another path and disown their child. Like so many other progressive offspring of bourgeois thought, the value and effectiveness of science itself, at least applied to society, has been called into question.

127

So, for example, the economist and sociologist F.A. Hayek, in a book appropriately titled *The Counter-Revolution of Science*, maintains that it was a mistake to attempt to apply the general approaches of the natural sciences to society, that a "counter-revolution" in science is now needed to rectify the fault. And recent histories of science have even tried to re-write the history of the great bourgeois natural scientific achievements in line with more sceptical thought. For instance, using dubious scholarship, some of these histories hold that Copernicus did not really think that the sun is the centre of the solar system: His theory to this effect was just a convenient fiction, not purporting to represent things as they really are.

Given this sort of anti-objectivist attitude, it is not surprising that classic Marxist works of philosophy, such as Frederick Engels' *Anti-Dühring* and V.I. Lenin's *Materialism and Empirio-Criticism*, are devoted in large part to defending objectivism. If the anti-objectivists were right, then Marxist writings about the contradictions of capitalism, the nature of imperialism, the need for a revolutionary political party, and so on would not be theories recommended for rational evaluation, but castles in the air. Neither Marxism nor any rival social theory could be used to arrive at true conclusions.

The attitude of anti-objectivism is so pervasive in North American academic circles that some who are otherwise attracted to the views of Marx are reluctant to accept the objectivist position explicitly advanced by Engels and Lenin. It is in response to this situation, I think, that several popular Marx "scholars" have maintained that whereas Engels and Lenin may have espoused objectivism, Marx himself was a foe of the view that objectivity is possible (and desirable) and that he substituted practice (or "praxis") for knowledge.

Since this view has proven an obstacle to many in trying to understand the Marxist position, it will be appropriate briefly to indicate its weaknesses:

1. If Marx really did hold the anti-objectivist position attributed to him, then, given the weakness of this position, I think we would have to say "so much the worse for Marx."

2. There is no good evidence, however, that Marx *did* hold such a view (those claiming he did usually misreading his works), while there *is* evidence that Marx held the same position as Engels and Lenin. Thus, Marx once praised Ricardo (see *On Malthus*) for sometimes coming to "scientifically honest" and "scientifically necessary" conclusions, even though they were at variance with the class position favored by Ricardo; and in a preface to *Capital* he holds that some social conditions are conducive to objective inquiry, while others are not. More-

128

over, there was a division of labour between Marx and Engels in which the latter usually addressed philosophical problems, and if Marx were at variance with Engels on this crucial philosophical point, he would certainly have taken issue with him, either publically or in their correspondence.

3. Marx did often emphasize the important connection between (objective) knowledge, especially that embodied in the theories of historical materialism, and practice, as did both Engels and Lenin. But rather than posing them as alternatives, he stressed their necessary unity. Only in practice can theories be developed and tested, and without objective knowledge success in practice could only be accidental.

4. The charge that Marx was opposed to the possibility of objective theories is often linked to the view that he was "dynamic" instead of simplistic and mechanical. This confuses two different issues. Marx was indeed opposed to mechanical and simplistic thinking, but so were Engels and Lenin, as should be clear to anyone who reads their criticisms of Dühring and the positivists (Empirio-Critics) precisely for being mechanical and simplistic. Marx, Engels, and Lenin alike were opposed to mechanical thinking just because such thinking fails to discover objective truths about society and nature.

What, then, is wrong with the anti-objectivist position regarding Marxism, or any other social theory of similar scope for that matter? I shall discuss five of the most common arguments.

Philosophical Liberalism and the Morality Argument

One line of argument, which might be called "philosophical liberalism" actually attacks *all* science, social and otherwise. It is urged that there may be a plurality of theories about any subject matter, each based on a different world view, or basic presupposition, or model, and that while "objective" agreements might be possible between people with the same world view, there is no way that disagreements which stem from having different world views can be settled. That is, there is no way to know that one world view is objectively more correct than another.

Another line of argument is reminiscent of the anti-bourgeois feudal attacks on the morality of science. Such is the view, for example, of many existentialists. The following is typical:

> . . . (with social science) comes the habit of looking upon man as we look upon the other animals, simultaneously failing to realize the consequences

of our change of attitude. But we cannot maintain our self-respect if implicit in our approach to ourselves and others lies the conviction that man is only that which science says he is. And one of the consequences of a loss of self-respect is that the fiber of actual living is coarsened. (E. Vivas, in H. Schoeck, ed., *Scientism and Values*: New York, 1960)

I do not think we need be detained by these two arguments. The first, though subtly presented and defended by many contemporary philosophers, is really a version of scepticism—the view, impossible for anyone to hold sincerely in practice, that no belief whatsoever can be justified. On this philosophical liberalism theory, for any belief which those who hold it regard as true and justified, there might be some world view in which the very same belief is false and unjustified—and there would be no way to decide which world view is correct. Each variety of this theory, whether based on particular philosophies of perception or language or whatever, has serious problems, but insofar as they have this sceptical consequence, I think that Engels' remark is appropriate of all of them:

> . . . taken in the abstract it sounds quite sensible. But suppose one applies it. What would one think of a zoologist who said: A dog *seems* to have four legs, but we do not know whether in reality it has four million legs or none at all? Or of a mathematician who first of all defines a triangle as having three sides, and then declares that he does not know whether it might not have 25? That 2 x 2 *seems* to be 4? (In *Dialectics of Nature*, Notes.)

As for the morality argument, perhaps two brief comments might be in order. First, science is a tool, and like any tool, it may be used in more than one way. Social science has indeed been used for immoral purposes, especially in the service of capitalists, but this does not mean that it is necessarily an immoral tool. Secondly, it is doubtful that the author of the passage quoted had encountered a wide variety of social-scientific theories about human nature. A mark of *bourgeois* theories is that they tend either to make humans into saints or brutes, and in the latter case (also in my opinion in the former) if those theories were the *only* social-scientific ones, then social science *would* lead to a loss of self respect.

There are, however, three arguments currently used against Marxism which bear more lengthy consideration. These are of special interest partly because they all contain a grain of truth, and partly because the views they are based on are both stated in *opposition* to Marxism by anti-scientific, anti-Marxists and *attributed* to Marx by pro-scientific, anti-Marxists.

130

Human Consciousness

The first argument has to do with human consciousness. Anti-scientific critics of Marxism advance it this way: Marxism maintains that there are general laws of history just as there are general laws of nature. But there is a difference: Unlike inanimate objects and plants, humans have consciousness and through awareness of hypothesized laws and predictions can thwart them; e.g., a capitalist could read Marx's *Capital* and take steps to prevent the demise predicted for him there. Now Marx certainly did allow for the role of consciousness in history (which leads pro-scientific *fatalists* to attack Marxism from the opposite direction here), but this admission does not mean there are no laws to be found or predictions to be made.

First, not all social-scientific predictions *can* be thwarted whether or not people become conscious of them or want to change them. For instance, at a certain point the fate of the old feudal ruling classes in England and France was sealed, and the best their ideologists could do was despondently (or masochistically) watch their class's demise and bewail the loss of the good old days. No amount of conscious manipulation could save their ruling position: The upcoming bourgeoisie was too strong, and the contradictions of the old rule too glaring. The case is similar, Marxists hold, with the bourgeoisie today. Short of nuclear annihilation, defeat of that class by the proletariat can only be set back and made more difficult by capitalist use of social-scientific knowledge. Humans are not gods.

Secondly, even with the alteration of behaviour in accord with social-scientific knowledge, the general movements of history—the sum total of the actions of individuals—can be scientifically studied. And, as part of this study, the facts of people's desiring the alterations that they *do* make need to be explained by finding social-scientific laws. Why for instance, do capitalists try to alter the future as regards *profit* more than as regards pollution? Engels makes this point:

> Men make their own history, whatever its outcome may be, in that each person follows his own consciously desired end, and it is precisely the resultant of these many wills operating in different directions and of their manifold effects upon the outer world that constitutes history. Thus it is also a question of what the many individuals desire. The will is determined by passion or deliberation. But the levers which immediately determine passion or deliberation are of very different kinds. Partly they may be external objects, partly ideal motives, ambition, "enthusiasm for truth and justice," personal hatred or even purely individual whims of all kinds. But, on the one hand, we have seen that the many individual wills active in history for the most part produce results quite other than those they

131

intended—often quite the opposite; their motives therefore in relation to the total result are likewise of only secondary significance. On the other hand, the further question arises: What driving forces in turn stand behind these motives? What are the historical causes which transform themselves into these motives in the brains of the actors? (In *Ludwig Feuerbach and the End of Classical German Philosophy*, Section IV.)

Finally, it should be noted that this whole argument presupposes what it is trying to disprove—that there can be social laws. The argument again is that having discovered a law of society, a person can act so as to prevent predictions based on the law from coming about. But what does this mean except that the person can himself *predict* that if he acts in a certain way, he can prevent certain things from happening. Another way of putting this point is to say that the original law assumes that there is no conscious interference in some course of events. Once there is interference, then the conditions assumed by the law no longer hold and we have a new situation, one in which people are consciously intervening in the course of events: There is no reason why this *new* situation cannot be subject to scientific law. Indeed, many of the general laws of Marxism are of just this sort. Predictions of proletarian revolutions refer to the conscious activity of members of the working class and their allies to stop and reverse the processes of social decay fomented by capitalism.

Bias

A second argument against Marxism as a social science has to do with the biases of social scientists. Social scientists, it is argued, being human, have emotional and often political stakes in the outcome of their studies, and this necessarily clouds all their so-called objective work. Now Marxists are among the first to agree that ivory tower impartiality is a myth, and Marx was quite good at uncovering far-reaching class biases in even the most apparently apolitical theories. (Hence, again, some pro-scientific objectivists also use this argument against Marxism.) The mere fact of class bias (or any other kind), however, does not by itself disprove the possibility of objective social-scientific inquiry.

In the first place, there is a difference between being *biased*, that is, so blindly or stubbornly committed to a position that you cannot objectively evaluate it, and being *partial*, that is, wanting scientific research to serve some particular interests. Bourgeois theorists often suppose that only social scientists who had no class or other interests at all, who did not care about the social consequences of their scientific work, could be objective. It is incorrect, however, to equate being

132

biased with being partial. A person can be concerned about the social and personal effects of an investigation and be unbiased in pursuing it.

In fact, depending on the class position people have and the stage of their struggle, it is often *because* they are partial to advancing the interests of their class that they are objective. For instance, the early bourgeois theorists could see that unlike their feudal opponents, whose ideologists were experts at obscuring the truth (rivaled perhaps only by *present day* bourgeois social theorists), their class had nothing to lose and everything to gain by discovering objective truths, and they were right. They *did* discover some objective truths and the knowledge *was* of great service to them in their ascendency.

The situation has changed, however: Now social-scientific objectivity is not generally to the advantage of the bourgeoisie, though it is to the advantage of the working class. It is not in the best interests of the bourgeoisie to promote widespread objective study of the general movements of history or of the nature of capitalist as opposed to socialist societies, for example. On the other hand, it is not in the best interests of the working class to allow bias to cloud objective truth (as working class movements have more than once painfully discovered). It might also be noted that to the extent that capitalism is becoming moribund and its educational systems more and more keyed, consciously and otherwise, to discouraging and fragmenting objective inquiry, especially in the social sciences, such inquiry becomes psychologically difficult for its apologists to carry out, even if they wanted to.

This bias argument also overlooks the particular nature of science as a mode of gaining knowledge. Scientific thought is characterized by its rigorous verification requirements. The whole apparatus of careful formulation of theories and predictions, of controlled experimentation and observation is designed to minimize the ill effects on an inquiry of the inquirer's prior views. Suppose that due to his class interests a social scientist says that monopolization of capitalist economies has had generally beneficial effects on the lives of the average working person, or that socialism is a hopeless ideal, since humans are essentially aggressive and selfish. Now regardless of his biases, either these views are in the main true or not, and it is surely possible to find this out. It might be hard work, not only the work of research, but theoretical and methodological hard work as well, e.g., defining "beneficial" or "aggressive" and deciding how broad a sampling is needed to provide significant data, what is to count as an exception, etc. But scientific thought *is* hard work, especially designed to arrive at objective truths, and this partly by eliminating the effects of bias.

The Relativity of Knowledge

The third general argument proceeds from the view that all knowledge is relative, i.e., is limited by the background knowledge, level of technology, interests, and general methodological techniques of the scientist's historical circumstances. According to some anti-scientific thinkers, this means that all truth is relative in the sense that what is generally considered true from the perspective of one historical period may be considered false from that of another, and it is impossible objectively to evaluate the two perspectives.

This view, called "historicism," is typical of many nineteenth-century German Idealist philosophers and was revived by the sociologist, Karl Mannheim, especially in his influential book, *Ideology and Utopia*. But since Marxists hold that all knowledge arises out of concrete historical situations and that ways of gaining knowledge vary with different social orders, Marxism is often *also* attacked for being a form of historicism not only by historicists opposed to science, but also by pro-scientific theorists. Hence, the professional anti-Marxist, Karl Popper, in his book *The Poverty of Historicism*, lumps Marx and Engels together with the German Idealists and attacks both for their relativism. (It should be noted that Popper, like other bourgeois critics of Marx, *also* attacks Marxism from a relativistic, anti-scientific perspective when it serves his purposes.)

Lenin devoted much of his *Materialism and Empirio-Criticism* to this very doctrine and concluded that it rested on a confusion:

> Two questions are obviously confused here: (1) Is there such a thing as objective truth, that is, can human ideas have a content that does not depend on a subject, that does not depend either on a human being or on humanity? (2) If so, can human ideas, which give expression to objective truth, express it all at one time, as a whole, unconditionally, absolutely, or only approximately, relatively? This second question is a question of the relation of absolute to relative truth. (In *Materialism and Empirio-Criticism*, Chapter 2, Section 5.)

The problem with historicism is that it poses a false alternative. If we cannot know everything all at once, if our knowledge is always incomplete and spread out in history, then, the historicist concludes, we cannot really know anything at all. Once the false dichotomy between knowing everything or nothing is abandoned, Lenin maintained, it is possible to put the sense in which knowledge is relative into proper perspective:

> The materialist dialectics of Marx and Engels certainly does contain relativism, but is not reducible to relativism, that is, it recognizes the relativity of all our knowledge, not in the sense of the denial of objective

truth, but in the sense of the historically conditional nature of the limits of the approximation of our knowledge of this truth. (In *Materialism and Empirio-Criticism,* Chapter 2, Section 5.)

The growth of scientific knowledge is relative because always limited by the techniques, available data, and so on of the scientist's times. Thus, no one historical period produces absolute knowledge, and the theories of different times and places may vary greatly. But this does not mean that there cannot be progress in the histories of the sciences, each age adding to the objective body of knowledge of preceding ones; and it does not mean that the dominant theories of any one time cannot be shown to be objectively false (for all times) at a later historical period. It has surely been shown, for example, that the flat earth theory is objectively false, even for those living in a time in which it was generally believed, and similarly with the once popular theory of the "invisible hand"—that if each individual were allowed to seek private profit in his own way, the whole society would run smoothly, as if guided by the invisible hand of God.

It might also be noted that this historicist argument, when pushed, becomes very much like the position above called philosophical liberalism: There are a variety of basic world views, each held in a different historical epoch, no one superior to any other, and as such, historicism has the same ultimately sceptical consequences noted by Engels.

There are many more issues involved in each of these arguments, but it seems to me that they form at least the foundation on which the most popular attacks against Marxism as a social science are built. By way of conclusion, it might be appropriate to note once again that many of these arguments leveled by bourgeois ideologists against Marxist social science do not attack it for being *wrong,* but for being *scientific,* that is, for claiming to be right. Clearly the torch of social science has passed from the theorists of the bourgeoisie to those of the proletariat—or rather it was taken away from the bourgeois ideologists before they could extinguish it!

The Relevance of Philosophy to Social Science

MARIO BUNGE

Introduction

It is well known that science and philosophy were one in the beginning and separated later on. It is less well known, but no less true, that they never lost contact but always interacted with variable intensity. This interaction has been asymmetrical however: Philosophy has contributed more—both good and bad—to science than the latter to the former. For one thing, philosophy handed over to science—though not in good grace—entire branches, among them psychology and social science. Secondly, all scientific research presupposes certain philosophical ideas, such as the logical principle of noncontradiction, the ontological hypothesis that there are laws, and the epistemological assumption that these laws can be known at least approximately. Thirdly, the philosopher has taken the trouble to examine the philosophical presuppositions taken for granted in scientific work. In sum, philosophy has made its mark on science.

On the other hand, the contribution of science to philosophy has not been impressive, and for this philosophers as well as scientists are to blame. There are philosophical schools, such as phenomenology and linguistic philosophy, that are uninterested in science, and others, such as scholasticism and existentialism, that are hostile to it. There is, then, more philosophy in science than science in philosophy, mainly because scientific research rests on philosophical presuppositions, whereas many people philosophize without taking any notice of science. This is the actual situation, not the desirable one.

I submit that a vigorous and symmetrical interaction between science and philosophy is desirable, to close the gap between the two camps and to develop a scientific philosophy and a science with a philosophical awareness. Such an ideal will not be attained by preaching philosophical sermons to the scientists, but by making an effort to understand them. I suggest that philosophers should become apprentices rather than lawgivers, and participants rather than onlookers. After all, scientists and technologists wield more power than humanists. Even if they were less arrogant they still would not listen to a philosopher unwilling to learn the ways of scientists and technologists—what their problems are; how they go about posing and solving them; and how they evaluate their own performance. It is only by changing our attitude, by becoming students of science and if possible part-time scientists, that we may hope to show scientists and technologists that they use and even produce philosophy—and that they should be able to learn something from the professional philosopher.

The first task in the endeavor to reunite science and philosophy is to exhibit the impact of philosophy upon science. This impact will not be easily discerned if one looks for philosophical terms and names of philosophers in scientific writings: Most of the philosophy in science is tacit. The impact will not even be noticed if one analyzes the explicit philosophical opinions of famous scientists on the nature and value of data and hypotheses, or any other epistemological theme, since it is rather common to profess a certain philosophical faith while practising another. The philosophical components are to be found beyond the phraseology, namely, in the mode of work, be it empirical or theoretical.

The impact of philosophy on science can be detected everywhere, but more clearly in the young disciplines in search of guidance or models. The guides are, of course, the older, established sciences on the one hand, and the philosophy fashionable among scientists on the other. Thus, the molecular biologist not only adopts the scientific method born in physics and chemistry but also, often, the mechanistic ontology that claims that the living is thoroughly reducible to the nonliving; to this end he identifies the organism with the aggregate of its chemical components. Similarly, the sociologist sometimes tries to explain social facts, such as imitation, from analogy with physical processes such as diffusion or magnetization, refusing to acknowledge that there is anything specifically social. On the other hand, most psychologists and even neurophysiologists work under the aegis of the body-soul duality invented by philosophers a couple of millenia ago. In all three cases, the philosophical influence is as patent as it is ambivalent. (For further examples of the ontology inherent in scientific research see Burtt 1932, Agassi 1964, Bunge 1967, Buchdahl 1969, or Bunge 1973a.)

There is little doubt, then, that philosophy exerts a strong influence on scientific research, from the choice of approach to the formulation of hypotheses and theories, as well as in the evaluation of all these components and of empirical data. This influence is sometimes positive and constructive, sometimes negative or inhibiting, and at other times ambivalent. For example, idealism—the doctrine that ideas exist by themselves—has been fertile in mathematics but lethal in physics. Its opposite, materialism, has been paralyzing in mathematics but stimulating in anthropology. And dualism—the view that body and mind exist side by side—was initially favorable to biological

137

research, by promoting the mechanistic approach, but has delayed the advance of psychology by divorcing it from biology.

Since the influence exists, the scientist may try one of the following strategies: (1) to discard philosophy, (2) to accept dogmatically the philosophy of the day, or (3) to filter the philosophical input retaining only those likely to be beneficial to science. The first strategy is, of course, the one most research workers pay lip service to. It is impracticable because every research project presupposes ordinary logic (which philosophy shares with mathematics) and has ontological and epistemological presuppositions, such as those of the autonomy and knowability of reality. The second strategy, though tacitly practised by many, is beneath the scientific spirit, which recognizes no idea as being beyond criticism. We are then left with the third strategy, that of filtering out the negative philosophical input into scientific research.

It is not easy to implement the filtering strategy, however, (1) because we do not yet have a well-developed philosophy totally sound and compatible with science, let alone unequivocally useful to it, and (2) because, even if such a philosophy were available, it would not always be easy to recognize or to know in advance which ingredients might have a beneficial effect upon science at a given moment. (Remember that some philosophical ideas are ambivalent or double edged, and that it would be disastrous, for science as well as philosophy, to adopt officially a given philosophical doctrine, declaring it, alone, favorable to the advancement of science.)

Nevertheless, only the filtration strategy is compatible with the critical approach inherent in scientific work and that may lead to the reunion of science with philosophy. But it behoves us philosophers to show scientists that this strategy—that of trying out selected philosophical ideas—can pay off. In particular, it is we who must show by example that we are capable of contributing to the advancement of social science, by alerting, clarifying, criticizing, and occasionally constructing. In the present paper I shall argue that we can help in locating problems, refining approaches, elucidating concepts, examining assumptions, organizing theories, recognizing laws, evaluating data, and even in characterizing social science and exhibiting its place within the contemporary cultural system.

Problems

Search and Evaluation of Problems

Scientific research starts by identifying and evaluating problems, neither of which is done in a conceptual vacuum. If we have no general orientation, we shall not even know what to look for. And if our general orientation is narrow, we shall tend to look for small problems. Thus, those who approach economics from a psychological viewpoint will tend to restrict their work to investigating consumer behavior which is part of market research. (See, e.g., Katona 1975.) Surely, the psychological component is important in certain sectors of the economy, such as the car industry in which exhibitionism and fashion prevail over economy and safety. But consumer psychology is incapable of explaining, say, the current world food crisis, initiated by a quick sequence of poor crops and the resulting exhaustion of stocks. In short, psychoeconomics is incapable of noticing, and *a fortiori* of solving, problems where individual psychology has little if anything to say.

Surely, the economist does not need the philosopher to realize that psychoeconomics is one-sided and therefore limited—if that economist happens to entertain a comprehensive conception of the social world—i.e., an ontology (even sketchy) of society and its subsystems. Whoever possesses such a comprehensive view understands that natural and human resources are more important than preference for this or that automobile make; that social structure is deeper than the manifestations or indicators of social status; that history—demographic, economic, social, and cultural—is more important than the succession of dynasties; and that the normal cases have more weight than deviations—except of course in the case of creations and inventions. The philosopher, precisely because he prefers general outlines over details and mistrusts partial views, may help recall all that. By the same token, he may help form a wide-encompassing and balanced view; hence he may help steer research towards significant problems, that is, questions whose answers elicit radical changes in our vision of society and our mode of controlling it.

Formulation of Problems

Once an important and presumably soluble problem has been found, the philosopher may help handle it at least in some phases of research. In fact he may:

(1) clarify the statement of the problem, which is often confused in social science, with the help of his general vision and his logical tools;

(2) help list the theoretical and empirical means necessary to tackle the problem concerned—means that may have to be created if they are not available;

(3) help recognize whether the proposed solution is genuine or just a heap of phrases couched in the fashionable social science jargon;

139

(4) help notice the logical consequences of accepting or rejecting the proposed solution.

At a time when many research projects cost huge human and material resources, the philosopher's collaboration may save millions of dollars—or may help waste them if wrong-headed.

Approaches

Fertile Approaches and Sterile Ones

Every investigation into an area of reality is done with a certain approach. An approach may be construed as a framework, a set of probems, and a set of methods—some general, others particular. The framework is the general orientation—for instance, the view of society as either a conglomerate of individuals (individualism), or a supra-individual totality (holism), or else a system with a definite composition (individuals) and a definite structure or organization (the relations and bonds among the individual components).

The approach will fashion the research programme and therefore the results of the investigation. For example, the individualist or elementarist will refuse to look for societal laws and the holist will be reluctant to search for the individual roots of collective action, whereas the systematist will be willing to embrace both aspects. The set of problems and the methods thus depend critically upon the approach. If the framework is consistent with the scientific spirit or outlook, then it will come together with the general method of science and the adaptations of the latter to specific problem areas, i.e., the techniques peculiar to the research field.

The philosopher may be useful at this stage of research by reminding the scientist that the results of any research depend more upon the approach than on the financial resources. An individual investigator possessing a fertile approach may be better than a large and well-heeled team of workers submerged in conceptual obscurity. (A penny for your thoughts—provided you have any.) The philosopher may make himself useful by taking an active part in the discussion concerning the various possible approaches. Let us recall two such polemics.

Two Central Controversies in Social Science

The two most important, or at any rate the loudest, controversies on the nature of social science are those concerning its framework and its method. The former divides the students of society and of social science into three groups: the individualists, the holists, and the

140

systematists. (For individualism and holism see the useful readings of Brodbeck 1968, Krimerman 1969, and O'Neill 1973. The systematic point of view is treated explicitly in Buckley 1967 and Bunge 1976, implicitly in the whole of mathematical sociology.)

In this regard the philosopher may be helpful in various ways: By pointing out, for example, that there have been similar controversies in other sciences, so that the social scientist may learn from them. (Recall the problem of microreduction in chemistry, namely the question whether the properties of molecules are somehow "contained" in those of the component atoms or are new, emergent, relative to the latter.) Another possible contribution of the philosopher consists in showing that the individualist (or elementaristic) approach, though clearer than the holistic one, is no less one-sided, and that both have actually been superseded by the systems point of view. The latter is fertile if only because it forces one to identify the components, the environment, and the structure of a system (e.g., a neighborhood) and because it suggests the search for subsystems (e.g., the educational system) and of super-systems (e.g., the regional economic bloc). But the philosopher should also warn against the fashionable illusion that a general theory of systems, devoid of specifically sociological (or economical or histori-cal) assumptions suffices to pose and solve particular problems con-cerning a given society.

As for the controversy on method, it has pitted the thinkers of the scientific school against those of the humanistic or traditionalist current. It boils down to the question: Is it possible to formulate mathematical models and to conduct experiments in social science? Certainly there is no need to be a philosopher in order to answer this. A mere perusal of the contemporary sociological literature will show the existence of mathematical models (e.g., Coleman 1964, Boudon 1967, Ziegler 1972, Fararo 1973, Alker *et al.* 1973, and the *Journal of Mathematical Sociology*) as well as of social experiments (e.g., Green-wood 1948, Riecken and Boruch 1974). The philosopher has a debt to pay to the social scientist, however, for in the past he succeeded in persuading him that social science looks more like literary criticism than like chemistry. To be forgiven for that fateful mistake the philosopher ought not only to recognize the existence of scientific (mathematical and experimental) sociology, but should also help it to develop, if only by dispelling the mistrust it still evokes in the traditionalist sector of social science. In particular the philosopher may show the great advantages in clarity, systematism, logical consistency, and deductive power gained by the use of mathematics.

Take, for example, the famous Parkinson's law, which ought to be an integral part of the theory of formal organizations. To endow it

141

with respectability, we shall formulate it thus: The efficiency of an organization increases with the number of its components until it reaches a maximum; from then on the efficiency declines until it becomes nought. At this point the size of the organization is twice the optimal size, and from then on the efficiency becomes negative—i.e., the organization consumes more than what it produces. The formalization is immediate. Call E the efficiency or yield of an organization, and N the number of its members. Then Parkinson's law reads:

$$E = k \ N \ (N_o - N/2)$$

where N_o is the optimal size (corresponding to maximal efficiency) and k a positive real number characteristic of the organization type. The meshing in of this formula with other statements in the theory of organizations and the empirical testing of it are as many challenges to the sociologist.

Concepts

Generic Concepts

Like every other science dealing with facts, social science employs certain generic concepts that it does not analyze: for examples, the ontological notions of entity (or thing), property, structure, change, novelty, and history; and the epistemological or methodological notions of hypothesis, law statement, theory, explanation, prediction, quantitation, measurement, and test. It behoves the philosopher to elucidate such concepts and to do it in the best possible manner, namely by building theories (ontological, epistemological, etc.) about them.

Take, for example, the concept of structure. Ask a structuralist, such as Lévi-Strauss, what he means by "structure": Chances are that the reply will be an elegant albeit unintelligible, hence useless, phrase. What will emerge with some clarity from the answer is that, according to structuralism: (a) the structure of a social group is not an objective property of the latter but rather of our own representation of it; and (b) structures are stable and oppose change. The philosopher may step in at this juncture, first critically, then constructively, and offer a clarification of the general notion of structure as a property of something, be it a set (as is the case with mathematical structures), be it a concrete system (like in every other case). Moreover, he may define the structure of X as the collection of relations among the components of X. And if he has any energy left, the philosopher may investigate the way the notion is actually used in social science (cf. Merton 1957, Blau 1974).

142

And if this investigation leaves him dissatisfied he may attempt to build an alternative concept of social structure, one that is mathematically transparent as well as utilizable in empirical research (cf. Bunge 1974b). But the problem of specific concepts deserves a section of its own.

Specific Concepts

Social science is characterized by a picturesque, pretentious jargon that often succeeds in hiding its concepts' lack of precision. It thus offers a visible target to the philosopher, who will be able to criticize conceptual imprecisions and even obscurities. But here, like elsewhere, his criticism will not be welcomed unless accompanied by constructive proposals (Bunge 1974b).

Assumptions

Generic Assumptions

For a time it was thought that science, whether natural or social, has no philosophical presuppositions: Genuine science is strictly "positive," i.e., a huge repository of data with no admixture of speculation. That belief turned out to be mistaken: We now understand not only that there is no science without theory, but also that the very search for data and building of theories proceeds in the light of a host of philosophical (logical, epistemological, ontological, and even ethical) assumptions. If this is true, if there are philosophical ideas in the background of every research project, who is better equipped than the philosopher to dig them out and examine them?

It does not take deep digging to uncover some of the ontological and epistemological presuppositions of every social science. For example, the very study of society as an entity existing independently of every investigator presupposes that society is a concrete object—just as concrete as a fish or a star—rather than a set of norms and values, and that, though just as imperceptible as an atom or a black hole, that entity is knowable to some extent. Another general assumption of a philosophical nature is that every society has properties—such as stability (or instability) and mobility (or rigidity)—characterizing it as a whole, properties that their individual components cannot have. A further philosophical hypothesis is that the value system adopted by a society is not an idea hovering above the components of the society but consists in the valuations effectively performed by them, and such that they are generally approved of or sanctioned by most members of the society. (A follower of Plato's or Hegel's would, on the other hand, regard a value system as an ideal object with an autonomous existence—which assumption would lead the social scientist to a wild goose chase.)

In addition to philosophical assumptions, all of the social sciences share certain generic assumptions that are not properly philosophical. For example, the hypotheses that (1) all societies possess certain cultural universals, (2) all societies resist change to some extent, and (3) this resistance is greater if the agents of change are external than if they are internal. Anthropologists and historians have confirmed these and many other cross cultural generalizations, but philosophers—with their knack for generality—are apt to notice further generic assumptions if they employ their analytical competence in examining the views discussed in the various social sciences.

Specific Assumptions

Just because nothing human is alien to him, and because he adopts a critical attitude, the philosopher is in a privileged position to uncover assumptions that the specialist may overlook and to evaluate hypotheses that the specialist takes for granted. For example the philosopher may wake up the statistician who compiles quality of life statistics convinced that his work is purely descriptive. He may be told that such a work involves norms as soon as he intends to find differences between actual data and desirable values or else thresholds: data on the percentage of the population of an area whose standard of living is below the poverty line, for example.

The social scientist is shy about his hypotheses, partly because of the empiricist tradition according to which hypothesizing is wicked, only data gathering being virtuous. And even if he exhibits some hypotheses, logical analysis is apt to discover that he has not formulated them all, or that they are not altogether clear, or that they are not well organized. A philosophical criticism may help remedy such shortcomings, or at least point them out.

At other times the philosopher must abandon his function as a critic and take the defense of certain social scientists. For example, data collectors dislike general assumptions and theories: They tend to believe that whatever is general is also abstract, hence remote from reality. In this case the philosopher may help by pointing out the difference between what is abstract or uninterpreted—such as a logical formula or a theorem in abstract algebra—and what is general, or referring to a wide class of items. The philosopher may thus save a good theory from narrow-minded attacks.

The social scientist may also welcome the philosopher's help when the former is accused of employing concepts that have not been given operational definitions. Here, the defense may show that actually there

144

are no such operational definitions. (See Bunge 1967, Ch. 3.) What scientists call "operational definitions" are often relations between observable and unobservable properties, particularly the indicators of economic activity, political stability, and cultural activity. The philosopher may wish to argue that no genuine scientific theory restricts itself to correlating observables, that theoretical concepts are not data packages, and that the verifiability requirement is met by just having the unobservables manifest themselves through indicators. (Think of medical symptoms as indicators of physiological malfunction or of infection, of behavioral phenomena as indicators of neural processes, and so on.) It should also be added that a reliable indicator is one backed by some theory, for only a theory can explain why a given variable is sensitive to certain other features. (See Bunge 1975.) But the subject of theory deserves another section.

Theories

Nature and Value of Theory

Data collectors distrust what they call "mere theories" and sometimes even oppose theory to (respectable) research. The philosopher can be of great help at this juncture by offering general considerations, such as: Every explanation and every prediction require theories, and the more exact, the greater the required exactness of the explanations and predictions. He may also point out that, far from being at odds, theory and genuine research (as opposed to aimless observation) are inseparable—that scientific research includes theorizing. Indeed, the search for data is always conducted in the light of some general orientation or conceptual framework which, if not a theory proper, should serve as a matrix for building theories. If the framework is poor or confused the compiling of data, however precise, may be a waste of time. In any case, why should one want to have data unless it is to motivate the construction of theories, or to put theories to the test, or to feed it into theories in order to explain or predict?

Social scientists usually prefer the word "model" to the term "theory." Every theoretical model (particularly if mathematical) of some area of reality is, however, a theory, i.e., a system of logically related propositions or formulas that form either a premise or a conclusion of a logical argument. To be sure, social science does not yet possess any theory as general and powerful as the fundamental theories of physics. But theoretical chemistry and theoretical biology, though older than social science, cannot compete with physics in this respect.

Of paramount importance in judging the achievements of a field of research is not so much the level attained as the rate of growth—and the latter is appreciable in all of the social sciences. For example, there are fairly exact and fertile models in econometrics, in the area of social mobility, in the study of organizations, and even in politology and history. They are certainly restricted to systems of rather narrow kinds and so to limited epochs, but they are theories nonetheless. Hence, there is no justification for the philosopher to deplore the theoretical poverty of the social scientists, especially since that poverty, now decreasing, was a consequence of an obscurantist philosophy that denied *a priori* the very possibility of building a science of society.

Theory Construction

The philosopher does not have recipes for building theories, but he can help construct (or destroy) them. First, he can and ought to pass judgment on whether everything that is called a "theory" in social science really is one. He will find that sometimes a mere taxonomic schema, at other times a chaotic heap of opinions, or at best a set of relevant but still programmatic utterances is called a theory. The philosopher must insist that the terms "theory" and "theoretical model" be reserved to designate hypothetical-deductive systems.

Secondly, although the philosopher can offer no infallible rules for inventing theories, he can propose a general strategy of theory construction which may be summed up in the following stages:

(1) selection of a few features (properties) of the objects of interest–namely those that look salient and promising;

(2) representation of each feature by a precise concept (e.g., a set or a function);

(3) formulation of relations among the resulting concepts—such as, "Set A is included in set B," or "The rate of change of function f is proportional to f itself," or "The probability of the transition of an individual from group A to group B is inversely proportional to the distance between A and B";

(4) logical organization of the resulting statements with a view to discovering what their implications are and what our commitments are if we adopt certain assumptions—i.e., conversion of the given set of statements into a theory (or theoretical model);

(5) confrontation of some propositions of the system with some empirical data—bits of information which may have to be sought after the theory has been built or sketched.

The philosopher may also have to insist that no theoretical model can possibly embrace all of the features of its referents: In the beginning we must discard some information and thus be prepared to end with an excessively simplified or idealized model. One may also

146

have to insist that a scientific theory contain law statements, and that these are usually not the stable regularities that emerge from an analysis of data, but rather represent patterns occult to direct observation.

Thirdly, if the philosopher as such is not equipped to propose new scientific theories, he certainly ought to help in improving the organization or structure of the existing theories—which are usually ill-organized even in the most advanced chapters of science. In particular he can try his hand at doing foundations of social science, by axiomatizing certain theories. For this task definite recipes can be formulated. (See Bunge 1973b.) On listing the presuppositions (in particular the logical and philosophical ones), the basic (or primitive) concepts, and the basic propositions or axioms (or postulates), the philosopher need not turn social scientist: He may confine himself to applying his technical tools to a body of knowledge drawn from social science. What is gained by axiomatizing? Greater clarity, systemicity, and verifiability.

Laws

Behavior Lines and Laws

In the natural sciences there is a clear distinction between a law and a behavior or evolution line, such as a trajectory. A given law usually embraces infinitely possible behavior lines: The latter differ by the circumstances not by the laws. This distinction is rarely if ever made in social science. For example, there is a tendency to regard behavior or evolution lines as laws. This confusion may result because the low theoretical level associated with the positivist or empiricist approach seeks regularities in the data—e.g., in time series—and discourages the invention of high level ideas that might explain the data.

The philosopher may effectively help distinguish laws from behavior lines by representing the state of the system of interest (e.g., a community) as a point in an abstract space whose every coordinate is a property of the system (or rather an attribute occurring in the theoretical model of the real system). As time goes on the representative point describes a trajectory in that space, called the *state space* (or phase space) of the system. Every such trajectory represents the evolution of the system, i.e., its behavior line. The precise trajectory will be determined by both the laws and the circumstances.

Social Laws

We agree then that every scientific law (or law statement) belongs to some theory, big or small. If a generalization does not belong to any

147

theory, then it may be an empirical generalization—i.e., one suggested or warranted by empirical data—but not a law statement. And if a theory contains no scientific laws then either it is not a theory proper (but perhaps a conceptual framework), or else an extremely general theory instead of a theory referring to a well-defined species of entities. (See Bunge 1973a for the kinds of theory with regard to degree of generality.) We also agree that it behoves the philosopher to help trace the distinction between laws and histories (or behavior lines), particularly in social science where it is as necessary as it is ignored.

It might be thought that the above distinction is not usually made in social science because there are hardly any generalizations, be it empirical or theoretical, concerning human societies. But this is not true. (For examples of social laws see, e.g., Berelson and Steiner 1964; for examples of politological laws see, e.g., Dahl 1971.) Since nothing persuades like example, let us recall a few generalizations found in various branches of social science:

(1) The birth rate in a community is directly proportional to the infant mortality and indirectly proportional to the standard of living in the same community. Practical corollary: To decrease the birth rate it is enough to raise the economic and sanitary levels.

(2) Social change is more frequent in heterogeneous than in homogeneous societies, and it is the deeper the more pronounced the stratification.

(3) The concentration of economic power is accompanied by a concentration of political and cultural power.

(4) The cohesiveness of a community depends on the active participation of its members in various groups or activities, and decreases with segregation. Cohesiveness also decreases with extreme participation, however. There is then an optimal degree of participation, corresponding to maximal cohesiveness, and that optimal value lies between full participation and no participation at all. Corollary: Both extreme centralism and extreme regionalism have disintegrating effects.

(5) Industrialization tends to destroy the extended family and favor the nuclear family.

These generalizations are actual or potential law statements. The conversion of an empirical generalization into a genuine law statement is greatly facilitated (though not insured) by casting it in precise mathematical terms. For example, generalization (5) gains in precision when formulated with the help of the statistical concept of linear correlation. But this will not suffice: The exact statement will have to

148

be incorporated into some (reasonably true) model of family organization. And in such a case it would be either a postulate or, preferably, a consequence of hypotheses concerning the mechanisms of formation and disintegration of the family in industrial societies. In fact, (5) can be explained by noting that industrialization promotes migration (internal or external), which in turn tends to disintegrate the peasant family (which is typically extended)—to which further negative factors are added such as the high cost of living and the dearth of housing in the urban communities which absorb the immigrants.

In sum, the philosopher can make himself useful by encouraging the social scientist to look for the laws underlying the histories, as well as by stimulating him to build theories housing such laws.

Data

Data Are Means

Many social scientists, misguided by the classical positivist framework—nowadays abandoned by all philosophers—collect data for the sake of data. The philosopher should criticize this cult of the raw datum by pointing out that, far from being an end, every set of data is a means—either to suggest theories, to activate them, or to correct social ills such as unemployment and the concentration of power.

Social scientists are fond of speaking of the "conclusions" they derive from empirical data. Surely a few low-level inductions can occasionally be found by examining a set of data—provided one has a good nose, experience, and luck. In addition, for such a lucky event to occur the data must be well presented, they must not be too "noisy" (i.e., affected by random variations), and they must in fact be grouped in a relatively simple manner. Such a conjunction of favorable circumstances pertaining to both the data and the investigator is the exception rather than the rule: The underlying laws are usually so complicated, and the data so "noisy," that direct generalization is impossible. In such a case, which is typical, the investigator will have to hazard hypotheses—which are not, of course, arbitrary or wholly speculative. All this, which should by now be obvious to the philosopher, is often ignored by the social scientist, still under the spell of inductivism. (For a definitive criticism of inductivism see Popper 1959.)

Another circumstance unfavorable to any direct generalization from empirical data is the characteristic ambiguity of social and historical data—something philosophers have still to learn. Take for instance the divorce rate in a given area: If low, shall we conclude that most couples are happily married or, on the contrary, that there is a strong social (cultural) pressure against divorce that causes a lot of unhappiness not recorded by the statistics? If, on the other hand, the divorce rate is high, shall we conclude that people do not much value marriage and family, or rather that they do value them so highly that, in order to save them, they do not hesitate to correct their own mistakes? Obviously, nothing can be concluded from a raw divorce rate without further investigation of the social, economic, and cultural context. And the same holds for many other of these statistics. (Think, for example, of the number of hospital beds, or of schools for the mentally retarded, or of jails.) The philosopher, who is temperamentally rather indifferent to data, may help the social scientist understand that data gathering is only a part of social science research.

Data Are Not Neutral

Another aspect of empirical research that may benefit from the participation of the philosopher is the selection of the kinds of data that are to be collected. Indeed, where the collector of data recommends beginning by gathering data and ending by asking oneself what they indicate, and where the speculative thinker limits himself to formulating his personal impressions with total contempt for the data, the critical philosopher is apt to suggest the following strategy:

The kind of data to be sought depends (a) on the approach that has been adopted and (b) on the available techniques and resources. It is convenient, therefore, to begin by exhibiting and clarifying the former and by weighing the latter. The philosopher as such may not be competent to make any pronouncements about the techniques and resources of empirical research, but he is presumably competent to discuss the matter of the approach. If he is an individualist or atomist, he will recommend that only data of a psychological type be collected; if he is a holist, that only global data be sought; and if he is a systematist, that data of both kinds be secured. Moreover, if he is a systematist, or is merely aware of what goes on in social science, he may limit himself to noting (a) that in fact data of both kinds are of interest and therefore sought, and (b) that many global data are aggregates (e.g., averages) of individual data, as is the case with the GNP or with average school education or with the dispersion of income.

In short, the philosopher is in a position to argue, not only that every datum is favorable or unfavorable to some theory, but also that every piece of information is collected and utilized in the light of some conceptual schema or scaffolding.

Place

The Unity of Social Science

Every specialist believes it his duty to avoid looking at other fields, even if contiguous. The philosopher, who is a generalist par excellence, may point out that this isolation is artificial and pernicious: that social reality is united, not fragmented, and that there is an optimal way of studying it—scientifically. It is not just that the central referents of all the social sciences are the same—people living in society and interacting with the physical environment—but that there are no exclusively economic, political, or cultural facts. Every fact embracing a social group, i.e., transcending the privacy of the individual, has those three aspects simultaneously—although one may be more pronounced than the others simply because every individual belongs simultaneously to the three main subsystems into which every society can be analyzed: the economy, the polity, and the culture.

Therefore, although we may and ought to focus our attention on one aspect at a time—for otherwise we cannot even get started—we must not ignore the others. How can one hope to understand, let us say, the cultural life of a community without knowing what it lives off, what its political régime and communications network are, and which its traditions and value system? It is one thing to temporarily close an eye as a methodological resort, i.e., in order to tackle circumscribed problems and build one-sided but precise models, and quite another thing to lack one eye—the eye for neighboring fields and comprehensive views. The philosopher may help the specialist keep his two eyes open—as in fact the social anthropologist tries to do all the time. And he may also contribute to integrating the partial results obtained by the cyclops in the various areas of social science.

For example, when dealing with the problem of choosing development indicators, the philosopher will refuse to adopt exclusively economic, political, or cultural indicators. He will probably insist that each of these is but a component of a vector that, taken as a whole, may be regarded as a measure of the level and rate of development; moreover, he may wish to equate the latter with the quality of life rather than, say, with industrialization, or modernization, or Americanization. (See Bunge 1974c.)

In conclusion, the philosopher will defend the unity of social science just as he will argue for the unity of physics (or of biology) beneath the divisions of labor.

Place of Social Science

Because he is a generalist the philosopher can help the social scientist locate his own discipline within the cultural system, and even place the culture as a whole within society. For example, he may point out that social science is part and parcel of factual (or empirical) science, which is in turn a part of intellectual culture. He may also note that further components of intellectual culture are technology, mathematics, and the humanities—including, of course, philosophy. The whole looks like a four leaf clover, with the leaves distinct yet united at the center and in turn united to the rest of society. It is the philosopher's responsibility to insist that the flows of ideas among the four leaves is indispensable to keep intellectual culture alive—no less necessary than their parity and the dependence of all four upon the society that nourishes and utilizes that culture.

Moreover, if he adopts a systems approach the philosopher may point out that social science does not exist by itself but is just the specific activity of the people who study society in a scientific manner. In fact, there are no ideas detached from those who think them nor, in particular, science without scientists nor, in general, culture without cultivators. The culture of a community is not a set of ideas, attitudes, norms, and values, but a system of persons who think, adopt attitudes and norms, and perform valuations.

In particular, let us insist, social science is a system of persons engaged in understanding society by learning from one another and from the rest of society, posing problems and trying to solve them with the help of whatever may look promising—not excluding biology, mathematics, and philosophy. It is only when analyzing and considering the merits of ideas independent of their personal and social matrix, that we are justified in (temporarily) detaching the scientific work (studying, investigating, training) from its products (problems, data, hypotheses, theories, techniques, recommendations) to the point of talking of ideas by themselves and of their influence upon society. Those who take ideas seriously do not degrade them to the condition of phantasmagorias but regard them as an aspect of the cultural process.

Conclusion

In the past the attitude of the philosopher towards the study of social systems and social events has been rather arrogant: He either declared that society is an unanalyzable, hence unexplainable, whole; or he thought himself equipped to study it. The latter attitude is found in the eccentric claim that sociology is a branch of the theory of knowledge. (See Winch 1958.) Nowadays, the philosopher is learning to take a new, more modest, and also more fruitful attitude: He has started to study some social science and to examine some of the philosophical problems it poses. This may be of help to the social scientist, particularly if the philosopher succeeds in drawing his attention to the philosophical ideas that are in fact employed in social science research. But the collaboration may go a step further by embracing—as we have suggested above—the discussion of approaches and concepts, hypotheses and theories, methods and the evaluation of problems and results.

It is certainly possible that the philosopher will not be heeded: The average social scientist prefers keeping his philosophical scaffolding to having it torn down by a philosopher well versed in contemporary philosophical techniques. Nevertheless, the probability of being listened to by the social scientist increases considerably if the philosopher becomes familiar with social science to the point of making original contributions to it, thus functioning himself as a living bridge between the two fields. Such a conversion is easier than it seems at first sight—at any rate easier than the transformation of the philosopher into particle physicist or molecular biologist—since even philosophers have some experience of social life and are informed to some extent of the current ideas about society and its evolution. Besides, every authentic philosopher takes an interest in everything and is in a position to discuss intelligently ideas of any kind—provided he takes some pains to understand them.

It would be desirable, along with having the philosopher draw nearer the social sciences, to persuade the social scientist of the convenience of studying some philosophy instead of abusing it or of clinging dogmatically to obsolete philosophical ideas. Moreover, it would be beneficial if every social science department and research institute and governmental agency counted its philosopher. (Think of the reduction in budget that can always be obtained by discarding projects for being muddled, immature, or superficial.) But such proposals will elicit understandable resistances unless the philosopher concerned is ready to learn to communicate with his scientific colleagues instead of locking himself up in his ivory tower. And to this purpose he will have to start by studying some contemporary social science instead of confining himself to reading and examining some of the classics—or, even worse, remaining in the grip of mere ideology.

To sum up, it is possible and feasible for philosophy and social science to reunite or at least to draw nearer each other and interact far more strongly than they do at present. Such a *rapprochement* should be fruitful for both disciplines—if only to alert the social scientist to the philosophical burden he is carrying anyway, and to engage the philosopher in a useful activity.

References

Agassi, Joseph (1964). "The nature of scientific problems and their roots in metaphysics." In M. Bunge, Ed., *The Critical Approach* (New York: The Free Press), pp. 189–211.

Alker, H.R., Jr., K.W. Deutsch, and A.H. Stoetzel, Eds. (1973). *Mathematical Approaches to Politics* (San Francisco: Jossey-Bass Inc., Publ.).

Berelson, Bernard, and Gary A. Steiner (1964). *Human Behavior: An Inventory of Scientific Findings* (New York: Harcourt, Brace & World, Inc.).

Blau, Peter (1974). Presidential address: Parameters of social structure. *American Sociological Review* 39: 615–635.

Boudon, Raymond (1967). *L'analyse mathématique des faits sociaux* (Paris: Plon).

Brodbeck, May (1968). *Readings in the Philosophy of the Social Sciences* (New York: The Macmillan Co.).

Buchdahl, Gerd (1969). *Metaphysics and the Philosophy of Science* (Oxford: Clarendon Press).

Bunge, Mario (1967). *Scientific Research*. Vol. I: *The Search for System*. Vol. II: *The Search for Truth* (New York: Springer-Verlag).

——— (1973a). *Method, Model and Matter* (Dordrecht: D. Reidel Publ. Co.).

——— (1973b). *Philosophy of Physics* (Dordrecht: D. Reidel Publ. Co.).

——— (1974a). "Les présupposés et les produits métaphysiques de la science et de la technique contemporaines." *Dialogue* 13: 444–453.

——— (1974b). "The concept of social structure." In W. Leinfellner and E. Köhler, Eds., *Developments in the Methodology of Social Science* (Dordrecht-Boston: D. Reidel Publ. Co.), pp. 115–215.

——— (1976). A systems concept of society. *Theory and Decision* 7:—.

Burtt, Edwin Arthur (1932). *The Metaphysical Foundations of Modern Physical Science*, rev. ed. (London: Routledge & Kegan Paul).

Coleman, James (1964). *Introduction to Mathematical Sociology* (New York: Free Press).

Dahl, Robert A. (1971). *Polyarchy* (New Haven and London: Yale U.P.).

Fararo, Thomas (1973). *Mathematical Sociology* (New York: John Wiley & Sons).

Greenwood, Ernest (1945). *Experimental Sociology: A Study in Method* (New York: King's Crown Press).

Katona, George (1975). *Psychological Economics* (New York: Elsevier).

Krimerman, Leonard I., Ed. (1969). *The Nature and Scope of Social Science* (New York: Appleton-Century-Crofts).

Merton, Robert K. (1957). *Social Theory and Social Structure* (Glencoe, Ill.: The Free Press).

O'Neill, John, Ed. (1973). *Modes of Individualism and Collectivism* (London: Heinemann).

Popper, Karl R. (1959). *The Logic of Scientific Discovery*. Translation of *Logik der Forschung*, 1935 (London: Hutchinsons).

Riecken, Henry W., and Robert F. Boruch (1974). *Social Experimentation* (New York: Academic Press).

Winch, Peter (1958). *The Idea of a Social Science* (London: Routledge & Kegan Paul).

Ziegler, Rolf (1972). *Theorie und Modell* (München: Oldenbourg).

Towards the Assimilation of Rules to Generalizations

ALEXANDER ROSENBERG

Certain opponents of the extension of methods from the natural sciences to the social have made a career claiming that the behaviour of human agents is rule-governed, and not regularity-governed, and therefore is different in kind from and not amenable to the same sort of treatment as the behaviour of entirely natural systems. The contemporary *locus classicus* of this view is Winch's *Idea of a Social Science* and several other works in the Routledge and Kegan Paul series in philosophical psychology. The work of these philosophers is characteristically unencumbered by actual examples of rules that contemporary social and behavioural scientists seriously employ in the explanation of the behaviour of human agents. In the present essay, I shall examine some actual examples of rules currently being employed in the explanation of behaviour. I hope to show that these rules bear no essential difference from relatively low-level generalizations in the natural sciences. I do this by asking and trying to answer for each of them why it is that the agents that follow them do so.

Agents, of course, do not merely follow rules; sometimes they *obey* them as well. The distinction between obeying and following has a venerable history, stretching back at least to Kant, who argued that we must distinguish morally valueless action which is merely consistent with the categorical imperative, from morally praiseworthy action, i.e., action undertaken in the light of, with regard to, out of the motive of obeying that rule. By way of arguing that rules are simply low-level generalizations, I shall also try to show that there is no important methodological or conceptual gulf that separates *obeying* a rule from merely *following* it. In both cases, the explanation of why an agent does it (follows or obeys) will turn out to be roughly the same.

It will be useful to begin with an expression drawn from the natural sciences, which is called a rule by physicists, but which I contend is clearly a low-level generalization. It will turn out that transformational rules of English accentuation advanced by Chomsky, the rules of kinship systems offered in structural ethnology following Levi-Strauss, and the rules I follow when I play chess bear striking affinities to this generalization and no methodologically important differences from it.

My strategy for encouraging the assimilation of rules to regularities is not the only one available. Another, perhaps more popular among philosophers, proceeds by offering a series of analyses of the complex of concepts at work in connection with rule-governed behaviour. A good example is Donald Davidson's "Actions, Reasons, and Causes," in *The Journal of Philosophy*, volume LX, 1963. Offering particularly important examples from the actual practice of social scientists is naturally complementary to this latter strategy. Moreover, it has the virtue of being neutral as between varying general accounts of the nature and role of rules. Whichever of these accounts is correct, there is good reason to believe that the finally accepted analysis should be consistent with findings about central and distinctive rules embodied in the best of current social research. To suppose otherwise evinces a revisionary view of the role of philosophy of the social sciences that seems unwarranted until we have a clear account of the actual character of social research. Moreover, it presupposes the enumeration of specific and irremediable defects in the best of current social science.

1. The Flux Rule in Electromagnetic Theory

Consider the following rule governing the electromotive force, the "emf" of a conducting electrical circuit:

> (I) The emf of a conducting circuit is equal to the rate at which the magnetic flux through such a circuit is changing.

The emf of a conducting circuit is the total accumulated force on the charge throughout the length of the loop. It may be suggested that (I) is not a rule at all and can bear no relevance to an elucidation of the relations between rules and agents who follow them. This suggestion has the character of a conclusion rather than a premise: Moreover, physicists call (I) the "Flux Rule," and it is not the only such expression which is called a rule in physics. There are also the right-hand and left-hand rules of electromagnetism, for example. Nevertheless, let us consider the reasons that might be advanced for this conclusion.

First, what are the subjects of the flux rule, the regularity of which system's behaviour is governed by (I)? The answer, obviously, is conducting circuits, any system that is capable of conducting an electrical charge. Although this restriction does not exclude agents (e.g., very sophisticated computors or robots), it plainly does not specifically envision them either. The rule connects the strength of the circuit's emf at any particular time with the rate of change of its magnetic flux at that time. That is, it connects two sorts of events or states of the system it governs. Clearly, conducting circuits do not *obey* (I), but why do conducting circuits *follow* this rule? The most natural

157

explanation is to show that the fact that they follow the rule is simply derivable from the laws of electromagnetic theory. Indeed, one is inclined to say that (I) is nothing but a rather low-level general law of that theory. Another way of answering the question is to cite the states referred to in the rule's antecedent and consequent and point out that the circuit is caused to have an emf of a certain strength by the rate at which the magnetic flux through the circuit is changing. More specifically, this rate provides a condition causally sufficient in the circumstances envisioned by (I) for the actual value of strength of the emf. In an influential paper on the nature of causation, "Causes and Conditions" (in the *American Philosophical Quarterly*, volume 2, 1965), J.L. Mackie has coined a technical term for such causal conditions: "inus" conditions, an acronym which expressed the claim that a cause is typically an *i*nsufficient but *n*ecessary component of a complex of conditions which was *u*nnecessary but *s*ufficient for the effect. Adopting Mackie's analysis, I shall assume that something is a cause if it can be shown to be at least a (contingent) inus condition. Naturally, on a regularity view of causation, to say that the rate of change of the magnetic flux is an inus condition for the strength of the emf is tantamount to the claim that the flux rule is derivable from the laws of electromagnetic theory.

Notice that (I), the flux rule, is not itself cited as causally responsible for the behaviour of conducting circuits. How could it be? The flux rule is not the sort of thing which could enter into causal relations. It is a rule, or at any rate, a proposition, and not a state, event or standing condition. What enters into the causal relation are items mentioned in the rule: the magnetic flux, its rate of change, the electromotive force, and its rate. Conducting circuits do not behave in a characteristic way *because* of the flux rule, any more than, as some philosophers have argued, agents engaged in rule-obeying behaviour are *caused* to behave thus by the rules. These philosophers have argued that rules which are followed or obeyed do not cause the relevant behaviour. They justify it, or provide its motive or reason, or bear some semantic or logical connection to the behaviour, making it understandable or reasonable—a connection plainly absent in the relations between conducting circuits and the flux rule. (Indeed, this is much the most pressing reason they offer for claiming that social rules are to be understood as causal regularities.)

In fact, the flux rule, and similar rules of physics, bear so little relation to their subjects, by comparison with, for example, chess rules, that one is tempted to say that if they are rules at all, they are not rules for conducting circuits, but for physicists dealing with such circuits. It was once popular among philosophers of science to talk of laws as

158

tickets or licenses of inference, and the flux rule may be a good example of what they had in mind. The flux rule is not a rule for the instruction of conducting circuits, but for the guidance of physicists. It licenses inferences from the emf to the rate of change of the magnetic flux. That it licenses inferences from the emf to the rate of change of the magnetic flux, however, is hardly sufficient to call it a rule, for all laws license inferences. So, to call (I) the flux *rule* really adds nothing to its content, and it is best viewed as a low-level generalization in electromagnetic theory. The question remains: Why do physicists call (I) and a few other expressions, "rules"? In many cases, I suspect, a general statement is called a rule simply because its earliest or most convenient expression makes reference either to the experimenter or to some instrument of his own construction. This seems to be the reason behind the so-called left-hand and right-hand rules, which make obvious, but unessential, reference to the experimenter's body. Even if the reference were essential, this would be no defect, nor lend credit to an anthropomorphic view of physical law, for surely parts of the body and the experimenter's instrument are subject to the same laws as other physical objects.

It is too soon to suggest that the flux rule bears no relevance to the rule-governed behaviour of agents, but we can conclude that, although physicists are entitled to their own linguistic conventions, the expressions they habitually call rules are innocent of any features that might distinguish them from mere regularities.

2. Rules of Transformational Grammar

Linguistic behaviour in all its complexity seems unique to humans (and perhaps some specially trained apes). Though highly complex, and apparently very disordered, linguistic behaviour has increasingly been understood as a rule-governed activity. Among the most impressive accounts of this behaviour are those of Noam Chomsky. In this section I shall examine one of the accounts that Chomsky has offered (together with George A. Miller) in "Introduction to the Formal Analysis of Natural Language," which appears in Luce, Bush and Galanter, *Handbook of Mathematical Psychology,* volume 2 (New York: Wiley, 1963). This is an account of the features of actual phonetic output by English speakers. Chomsky and Miller argue that in spite of its great complexity, actual phonetic output can be explained by "systematic cyclical application of a small number of transformation rules" to the syntactic structure of the utterance to be produced by the speaker. The syntactic structure of an utterance, its logical form, is determined independently of the phonetic structure of the utterance, its stresses and vowel reductions, by an analysis of, for example, the logical consequences of the sentence which the utterance expresses. Thus, the inscription "small boy's school" can mean "a school for small boys" or "a small school for boys." The meaning can be determined by an informed user of the inscription from the inferences surrounding the phrase. Of course, in spoken English we are in fact never confused about what "small boy's school" means, because phonetic stress conveys its meaning. Chomsky and Miller have provided rules which connect the differing syntactic structures and the differing phonetic outputs, and they argue that "it seems reasonable . . . to assume that rules of this kind underlie both the production and perception of actual speech."

The following list of rules has been adapted from Chomsky and Miller's account. When properly applied to the assumed underlying or so-called "deep" syntactic units (symbolized either in grammatical tree-networks or by nested parentheses), they correctly generate the accentuation of a large class of English words and phrases:

(a) A noun or stem in the first position of the syntactic unit is stressed.
(b) If a syntactic unit is enclosed with another larger unit the first principal stress of the smaller unit is dominant, and reduces the other stresses by one degree.
(c) If several principal stresses occur within a syntactical unit the last principal stress is dominant and lowers the others by one degree.
(d) A vowel is reduced if it does not receive a stress in the course of any cycle of transformations, or if the cycles of successive transformations have brought the principal stress which it carries to a third or sometimes a second level.

These "transformational" rules determine "the phonemic effects" of the constituents of the logical or syntactic structure of the proposition to be expressed. Thus the serial application of these rules to, for example, *small* (*boy's school*), i.e., small school for boys, results in the most pronounced stress figuring in the utterance at boy's; by contrast the application of these same rules to (*small boy's*) *school*, i.e., school for small boys, results in the first word, small, receiving the principal stress. Remarkably, the transformational rules not only determine phonetic features as a function of sentential syntax; they also determine stress, and reduction among the syllables of single words. Chomsky and Miller conclude that:

We have only a single cycle of transformation rules, which, by repeated application determines, the phonetic form of isolated words as well as complex phrases. The cycle ordering of these rules, in effect, determines the phonetic structure of its underlying elements. [These rules] are the basic elements of the transformational cycle in English.

Let us consider the extent to which these transformational rules, (a)–(d), differ from the flux rule. One obvious difference is the sorts of subjects to which (a)–(d) are applicable. They deal with regularities in the behaviour of English speakers, a set of items which has at least some members *not* in common with the class of conducting circuits. Moreover, it might be argued that insofar as these rules deal with English speakers they deal *ipso facto* with conscious (if not human) agents. But it is at this point that striking differences between (I) and (a)–(d) end. This is roughly because, although these rules govern aspects of the behaviour of agents, the fact that they are followed by agents is not to be explained by reference to any of the agents' *conscious* states, i.e., the states which characteristically distinguish agents from, say, conducting circuits. Although Chomsky and Miller explicitly cite (a)–(d) in order to explain regularities in the accentuation pattern of English speakers, they do not suggest that the rule and the behaviour are mediated by any conscious processes. Of course, they could hardly make such a claim, for introspection alone would immediately refute it. Moreover, they make it a virtue of their theory that it accounts for verbal behaviour without appealing to dubious conscious mechanisms. How then are we to explain that these transformational rules relating syntactic deep structure to phonetic behaviour are invariably followed? I suggest the answer to this question has the same conceptual character as it did in connection with conducting circuits and the flux rule.

Notice that in each case the rules (a)–(d) connect an antecedent, expressing the form of a syntactic deep structure, to a consequent, stipulating a level of stress for the associated phonetic unit. Now, that a proposition has a particular syntactic structure and that its associated English verbalization has a characteristic stress pattern are what Hume called, "separable existences." The connection between them is surely not of a logical sort. If, nonetheless, agents follow rules that constantly conjoin them, then there must be a causal relation between them. In particular rules (a)–(d) are followed because a proposition's having a certain syntactic form is at least an inus condition of the stress pattern of its vocal token. Because there is normally no conscious "recognition" of syntactic structures, nor intentional production of stress patterns, it is difficult to make out what the states or events are which constitute the realizations of the relevant causal antecedents. The rules (a)–(d) are not to be found in any guide for the correct pronunciation of English, either for natives or foreigners, but they can be programmed into a computer which will be able to produce an unambiguous simulation (or someday, perhaps, a duplication) of actual stress patterns for any given proposition. This will shed some light on our problem. Programming a computer to follow a rule like (a)–(d) is

161

simply changing the functional arrangement of the microcomponents of the computer, so that they will respond to initial conditions— inputs—in certain ways, which responses will aggregate to particular output. If the computer's operator introduces a proposition, the computer's following rules (a)-(d) consists in the analysis of the proposition's syntactic structure, and the production of a stress pattern for it, where (assuming the computer is relatively simple) there is no question of consciousness and the syntactic structure is plainly the inus condition of the stress pattern. The computer follows the transformational rules in just the same way and at the same time as it is following the flux rule; for in an electronic computer, of course, the rate of change of the magnetic flux in its conducting circuits is at least an inus condition of its emf. This suggests that English speakers' following (a)-(d) consists in the existence of an input-output system in a nonconscious area of the brain, which enables the (non-conscious) determination of syntactic structure to cause the non-conscious production of a specific stress pattern. (Notice that this view has no logical connection with central-state materialism, a claim about the identity of all conscious states and some brain-states.) The determination of syntactic structure is an event, and so is the determination of associated stress patterns. Both are typically non-conscious; their occurrence is not normally monitored or controlled by conscious centers, though these latter can and sometimes do "override" the causal forces involved. That is why syntactic determination is only at least an inus condition of accentuation.

It may be objected that I have completely failed to recognize that in humans, at least, behaviour in accordance with (a)-(d) presupposes that the agents are engaged in *speech*, and speech is pre-eminently a conscious, intentional activity. In consequence, any assimilation of the rules of transformational accentuation in English to rules like the flux rule are utterly unwarranted. This objection is both false and irrelevant. False, because at least some speech production that follows (a)-(d) is unconscious (e.g., talking in sleep). Irrelevant, because the rules of accentuation govern, not the conscious production of meaningful noise, but a *certain characteristic or property of that noise,* which is clearly not consciously produced. In general, it is important to distinguish between an event consciously produced, and the properties of that event which, though essential to the aim for which the event was produced, are not themselves consciously intended features of the event. A good example is the witch-doctor's administration of a cure of mythical repute which, unbeknownst to him, contains effective medication. Even closer to home, the normal decibel level of my speech is not normally consciously produced—i.e., before the mind—(except in

162

whispering or shouting) but the happy fact that it is within a certain level is essential to communication. Applying this distinction between consciously produced events and their essential but unintentional properties to the present case, it becomes clear that because the production of vocables generally requires consciousness is no more relevant to agents following transformational rules of accentuation than the conscious production of vocables by an agent with a prosthetic electronic larynx is relevent to his prosthesis following the flux rule when he talks.

The account I have offered is somewhat speculative, but I see no other reasonable alternative consistent both with the Chomsky-Miller view and the general canons of scientific theorizing. If the speculation is warranted, then the transformational rules of accentuation may someday themselves be explainable by reference to neurophysiological facts and events. In consequence, (a)-(d) may turn out to bear to an as yet unformulated theory the same relation that the flux rule bears to electromagnetic theory.

This at least appears to be the direction which the Chomsky-Miller approach must take, if they continue rightly to insist on advancing these rules as *explanations* of observed behaviour. There is, however, another path open. If this sort of elaboration of transformational linguistic theories encounters insurmountable obstacles, we may turn to viewing rules like (a)-(d) as rules or tickets of inference, tools of the "linguist," enabling him to predict as yet unobserved phonetic structures on the basis of a syntactic schematization, which, however, bears no immediate relation to the mechanism whereby linguistically competent agents produce these vocables. Even then, that it does permit correct prediction of unobserved accentuation would have to be explainable in principle by reference to some relation between the linguistic schematization it employs and the actual mechanism that determines accentuation. In other words, the rules of transformational grammar might some day come to have a status much akin to that of the right-hand rule briefly touched on above.

3. Structural Ethnology and Kinship Rules

Turn now to a set of rules which appear to bear even less affinity to the likes of (I) than (a)-(d) may initially have appeared to bear: The rules proposed by structural ethnologists, following Levi-Strauss, to subsume the diverse kinship rules that characterize a large number of primitive societies. According to one widely accessible version of this theory, presented by Kemney, Snell, and Thompson, in *Introduction to Finite Mathematics* (Englewood Cliffs: 1956), the following seven

163

rules characterize the marriage prescriptions and proscriptions of a great many primitive tribes throughout Micronesia and elsewhere:

(1) Each member of the society belongs to a marriage type.
(2) Two individuals can marry only if they belong to the same marriage type.
(3) The marriage type of an individual is determined exclusively by the sex of the individual and the marriage type of his or her parents.
(4) Two boys or two girls whose parents are of different types, are themselves of different types.
(5) Authorization or interdiction of marriage between two individuals depends only on the relations of their families.
(6) It is always possible for some descendants of any two individuals to marry.
(7) No one can marry his sister.

Consider the relationship of these rules to the marriage rules of two particular primitive tribes subsumed under (1)–(7): the Kariera and the Tarau. Although structural ethnologists insist that these rules are followed in the marriages of Kariera and Tarau tribesmen, no member of either tribe is acquainted with any of these rules save (7). Moreover, the rules which members of the former tribe do claim consciously to be following are *incompatible* with the rules which members of the latter tribe claim to follow. For instance, in Kariera society it is a rule that first cousins may marry, where the female is a daughter of a sister of the male's father or a daughter of the brother of the male's mother, while marriage is forbidden between a male and the daughter of his father's brother. On the other hand, Tarau marriage rules proscribe all cousin unions except marriage to the daughter of the male's mother's brother. These incompatible marriage rules can be generated from (1)–(7), by assigning different marriage-type transmission schedules to different tribes. For Kariera and Tarau tribes the following transmission schedules conjoined to the set of rules (1)–(7) generate the incompatible consciously articulated marriage rules:

	Marriage type of parent	type of son	type of daughter
Kariera	1	3	4
	2	4	3
	3	1	2
	4	2	1
Tarau	1	1	4
	2	2	1
	3	3	2
	4	4	3

164

As in the case of transformational rules of accentuation, there is no question here of agents consciously obeying rules (1)–(7). Nevertheless, the behaviour of agents can be explained by citing these rules and the transmission schedule. But, unlike the circumstances within which the theory of accentuation is offered, in this case there is for each tribe *another* set of rules which members of the tribe *can* articulate, which they describe themselves as following, and which can also explain this very same behaviour. What, then, is the relation between these two sets of rules, one of which is followed and the other of which is obeyed? Since both govern the same behaviour, it might be suggested that they operate independently and overdetermine its occurrence. Ethnologists certainly do not treat the behaviour in question as overdetermined, however, and the logical relations between the two sets of rules makes this suggestion at best implausible. Since conscious marriage rules are deducible from non-conscious kinship rules and the transmission schedule (which functions rather like a statement of initial conditions), it seems plausible to suppose that the conscious rules provide mutually incompatible "special cases" of the non-conscious ones, in the way that, for example, Euclidean and non-Euclidean geometries provide special cases of projective geometry. This view, however, seems to gloss over the apparently crucial difference between the two sorts of rules, a difference which cannot be attributed simply to the restricted character of conscious marriage rules. The latter operate through the mediation of consciousness, while the other sort do not appear to do so. We will return to this problem shortly.

Let us ask why it is that agents act in accordance with the structural ethnologist's rules. The answer will be roughly the same as that given in connection with the transformational rules of English accentuation. In the present case, the rules jointly associate "separate existences," marriage types, and marriage pre/proscriptions. The connections between types and pre/proscription are plainly not of a logical character. Since members of geographically and culturally diverse societies follow rules that constantly conjoin marriage types and pre/proscriptions, there must be a causal relation between them. In fact, we believe that rules (1)–(7) are followed by tribesmen because we have discovered that a tribesman's being in a certain marriage type is at least an inus condition of the marriages open to (or forced upon) him, and, therefore, is a causal condition of the marriage that he will eventually make. Again, there appears to be no question of these rules themselves or a knowledge of them being part of the deliberative process as a result of which the relevant behaviour is produced. On the other hand, unless we deny explanatory power to (1)–(7), there seems no alternative but to recognize a causal connection between items cited in the antecedents and consequences of these rules that accounts for

165

why the rules are followed. Moreover, like the transformational rules of accentuation, it should be at least in principle possible to show that kinship rules are themselves explainable by reference to some more general social or social-psychological theory. Accordingly, although they do deal with agents, they appear conceptually to be of a piece with transformational rules of accentuation and the flux rule of electromagnetism.

It might be objected that, unlike the transformational rules, kinship rules govern a sort of behaviour which is paradigmatically intentional and the product of conscious deliberation. Although this is true, it must be recalled that these rules do not cite facts which agents are conscious of, nor are agents conscious of following or obeying them. Thus, at this stage at least, that (1)–(7) govern behaviour which is considered by its agents to be intentional has no relation to the agents following them in roughly the same way that conducting circuits follow the flux rule.

But how can we reconcile that tribal marriage rules, which are consciously followed (and thus obeyed) are special cases, restrictions on the rules of kinship, and can be generated from these rules, with the conclusion that we have just reached to the effect that the rules structural ethnology offers us are not obeyed, but merely followed by agents no differently from the way conducting circuits follow the flux rule? Having ruled out overdeterminism in the explanation of the actual behaviour of tribesmen, we must either deny any explanatory force to the rules which the tribesmen cite to explain their own behaviour, or show that non-consciously acting in accordance with rules brought to light by ethnologists and consciously obeying the tribe's traditional dictums are one and the same thing. Although the first alternative has its attractions (and its exponents; Durkheim, for example), it seems unreasonable in view of the special/general relation evinced by the tribesman's rule and the ethnologist's one. Can we establish the second alternative? Can we show that the factors cited in (1)–(7) *cum* transformation schedule are the same as those which figure in the calculations of the agent obeying his conscious rules?

Fortunately, it is quite easy to show that an agent can obey one rule and simultaneously follow another without his action being overdetermined by the conjunction of these two rules. For example, a motorist who consciously obeys the rule "stop at all road signs bearing the word 'stop'" also follows the rule "Stop at all red hexagonal road signs," even if he has never noticed that all stop signs bear this shape and colour. More generally, if two rules R_1 and R_2 contain coreferential expressions, then if an agent obeys R_1 it does not follow that he obeys R_2, although it does follow that he acts in accordance with it. In other words, "A obeys R" is a referentially opaque context, while "A acts in accordance with or follows R" is a referentially transparent one. Notice that in the example offered above we can infer that the motorist follows one rule because he obeys the other, because there is something like a nomological or law-like connection between a sign's saying "stop" and its having a certain colour and shape. We need to add this restriction to our general claim about the logical relations between rules containing co-referential expressions: These expressions must describe types of behaviour that are nomologically connected. Return now to the tribesman's rule and the ethnologist's. It is clear how the tribesman can be simultaneously obeying one rule and following another without his action being overdetermined by the conjunction of the two rules. It has yet to be shown that his obeying one and following the other are one and the same thing, however. In the case of the motorist, it seems obvious that the rule he is merely following has no immediate role in the explanation of his behaviour; whereas that a wide variety of societies follow rules (1)–(7) lends considerable weight to the claim that they *do* explain tribesmen's behaviour, and, indeed, do so far more generally and systematically than the particular, consciously expressed, nomologically connected, co-referential rules of individual tribes. If we also accord explanatory power to these latter rules, then on the assumption that there is no overdetermination in operation, we have no alternative consistent with reasonable canons of causal inference but to allow that both sorts of rules make reference to the same sets of causally relevant items. In particular, the terms in which the different sorts of rules are couched should be understood as alternative functional characterizations of the same sets of items. The terms of the "versions" of the rules under which they are obeyed are couched in the functional language of ordinary discourse about blood relations. The terms of the ethnologist's version are couched in a functional language dependent on the ethnologist's characterization of kinship systems (which obviously is not the same as the primitive tribesman's). Both sets of terms pick out identical items which are causally responsible for the behaviour which the rules govern.

If in fact the structural ethnologist's rules, which are merely followed, are equivalent in the factors they cite to the rules which the tribesman obeys, then the role of these latter rules in the production of behaviour must be the same as the role of the former. But if the ethnologist's rules are related to the behaviour of tribesmen in the way that transformational rules of accentuation are related to English speakers, and in the way the flux rule is related to conducting circuits, then ultimately rules which are obeyed play the same role in the generation of behaviour as those which are merely followed: a role

167

which can be explained by reference to exclusively causal notions and for which it cannot be claimed that insofar as social behaviour is rule-governed, it is intrinsically different from and incommensurable with the behaviour of entirely natural phenomena. This is not to say that there is no difference between obeying a rule and following it, but only that this difference will not support the views that anti-behaviourists often try to rest on it. In the next section, I shall examine what that difference is.

4. Obeying a Rule of Chess

Whenever I play chess, my actions are governed by certain rules, among them the rule that

(K) The King must be moved out of check.

Unlike conducting circuits which only follow and never obey the flux rule, I can forebear from following this rule, and when I do follow it I am invariably also obeying it. Nevertheless, there are important similarities between the flux rule and (K).

What are the subjects of the rules of chess? Agents? No, for there are machines which are not agents and yet which play a correct if pedestrian game of chess. Why does (K) correctly describe the behaviour of chess players? Obviously, because players manifest a certain regularity: Whenever their kings are in check, they move them out of it. Among chess players playing their game, the king's being in check is at least an inus condition for the king's being moved out of check. And (K)—in so far as it is a rule chess players act in accordance with, merely follow—is correctly said to be followed simply because of this relation of inus-conditionality. Moreover, in the case of chess-playing machines, a well-articulated theory is at least in principle available to explain why these machines act in accordance with (K). Indeed, if we follow out the reasoning of the exponents of psychological explanation by machine simulation, such simulations, if rich enough, are *ipso facto* explanations of human chess-players' decisions as well. The availability of such an explanation might, in the case of machines, lead us to suggest that (K) is not really relevant to the performance of the system, and merely serves as a useful descriptive tool for those observing the behaviour of the machine, like the left-hand and right-hand rules in electromagnetic theory. But it would probably be more reasonable to say that (K) acquires explanatory force for chess players of the machine type from the existence of a theory which generates (K) as a regularity, just as the flux rule is generated as a regularity. We may

168

conclude that so far as machine chess players are concerned, (K), like (I), and the other rules we have canvassed, functions rather like a low-level generalization.

But suppose we say that chess players behave in accordance with (K) because they obey it or act in the light of it or because of it. We must say this for human chess players, both because we have no elaborated theory at hand, as we do for computers, and, more importantly, because human chess players themselves cite (K) as the conscious reason for their behaviour. The difference between computers that merely follow (K) and human chess players that obey (K) is brought out in the differences between the explanations of why both sorts of systems *follow* (K). For the machine, being in check is an inus condition for being moved. This is also true for humans; but there is an intermediary between the former state and the latter event, which is of a sort not normally supposed to obtain among machines, and which explains why the state and the event in question are related by inus-conditionality: The king's being in check is an inus condition for its being moved *because* (among other things) the chess player is conscious of rule (K). Under the condition that (among other things) his king is in check, consciousness of (K) is at least an inus condition for the king's being moved. Thus, when an agent obeys a rule, the causal mechanism involved is the same as that involved in his merely following it—except that one of the operative conditions is his awareness of the rule that he is following. To say that someone is obeying a rule is simply to cite a particular link in the causal chain between the items cited in the antecedent of the rule and those cited by the consequent by which the agent is *following* the rule. That this extra link is a conscious state is all that distinguishes obeying a rule from merely following it. But, in so far as followed rules appear very little different from low-level generalizations of science in their character and their relation to the behaviour they govern, there seems no reason not to draw the same conclusion about obeyed rules.

5. Conclusion

The conclusion that I hope to have reached is that there is no methodological or conceptual difference of any real importance between the low-level generalization of natural science that the flux rule exemplifies and rules that agents obey and/or follow: There are no differences unless the occurrence of a mental event in the causal chain involved in obeying a rule is *ipso facto* grounds for such a difference. But this view is so obscurantist that it is hardly worth treatment.

A more serious objection to this claim is that human agents can

169

and do often or occasionally break the rules which they are said to follow, whereas conducting circuits cannot do so. In other words, unlike even low-level generalizations, rules can be broken by the systems whose behaviour they apparently govern. With concentration I can change my patterns of accentuation; a tribesman contaminated by modernization can marry whom he pleases; Fisher can refuse to move his king and forfeit the game. Consequently, the objection goes, rules can hardly be considered low-level generalizations. Unfortunately, such generalizations are not always true. Consider the very low-level generalization,

(G)   Wood floats on water; iron sinks in it.

This general statement is, of course, false. There are types of wood that do not float and configurations of iron that will. The deficiencies of such a statement can, of course, be remedied, however, by attributing another property, density, to wood and iron which is connected to their behaviour in water and which enables us to produce other more general statements that serve to explain exception-riddled statements like (G). I believe that to some extent the movement from obeyed tribal marriage rules to the structural ethnologist's more general formulations mirrors this same kind of theoretical progress. Another way in which flaws like that of (G) and social rules generally can be remedied is by restricting the subject of the generalization, or by making reference to the ever-present but implicit *ceteris paribus* qualification which preserves so many of our most cherished generalizations from regular disconfirmation.

Now, there is no reason why the same preserving strategies cannot also be applied to explain why rules are sometimes disobeyed or not followed. In particular, an agent's conscious decision not to obey a rule should be understood as a result of the occurrence of factors which interfere with the operation of the rule in the way that a nail's being hammered to the barn interferes with its being displaced toward a magnet held in its vicinity. This means, of course, that for many scientific purposes rules will not be very useful, since they are shot through with exceptions and imprecisions, but this is already the case for low-level generalizations, and so cannot militate against our conclusion.

There remains one apparently essential feature of rules which I have not mentioned at all, and which many will insist provides a condition sufficient to distinguish utterly between rules and regularities: the element of prescriptiveness. Although the flux rule involves no prescriptive element, both the structural linguist's and the ethnologist's rules do manifest this feature. The rules of stress are supposed to do so

170

partly because, having internalized them, agents are able to recognize errors in stress committed by speakers and to correct these deviations, in spite of not being able to rehearse the rules themselves. The structural ethnologist's rules bear at least an indirect prescriptive force, since the articulated marriage rules of a tribe (which can be generated from them) plainly involve prescriptive sanctions. I admit that the element of prescription adds a complication to the character of rules which is absent among (other) low-level generalizations. Nevertheless, that an agent's following and obeying these rules can be explained in a purely causal idiom shows that any simple inference from the prescriptiveness of rules to the inapplicability of causal explanations is unwarranted. I suspect that it is simply because this feature is absent from most causal regularities that some philosophers have been led to draw a contra-causal conclusion about the character of rules. For example, Louch argues this way in *Explanation and Human Action* (Oxford: Blackwell's, 1966). To attribute prescriptive force to a rule *adds* something to its other characteristics without requiring us to delete any of those features which make it a regularity (as well as a rule). Moreover, I suspect that the attribution of prescriptive force is, in fact, to be understood by reference to characteristic sorts of causal relations. This claim in itself merits a book and not a paragraph, especially since there is as yet no widespread agreement on the meaning of the term prescriptive.

Let me, nevertheless, suggest the following: A rule is called prescriptive if, when the behaviour of an agent fails to follow it, this event is at least an inus condition for certain other sorts of behaviour—often verbal, but sometimes more physical—on the part of *other* agents. At least part of what is involved in the prescriptive force of rules is the existence of a mechanism which is, so to speak, triggered by rule-violation, and which acts to prevent future failures to follow the rule in question, or imposes some other constraint, condition, or sanction on the violating agent. Naturally, this suggestion comes nowhere close to exhausting the nature of the prescriptive force of rules, but it shows that we can understand this force on the model which we use to explain (typically human) responses to circumstances in which a regularity has failed to operate. Such circumstances cause us *to do* something: A boat-builder who finds that his wood doesn't float will (be caused to) find some other sort of wood; a computer's failure to follow a rule may call forth a replacement of one its components. I suggest that the only reason we do not attribute prescriptive force to regularities whose exceptions, too, call forth such "remedial" activity is that these rules do not govern human agents. But this will turn out to be no surprise.

Decision Making in Committees

ALEX C. MICHALOS

I. Introduction

Breathes there a committeeman with soul so dead, who never to himself hath said, "I could make this decision twice as well in half the time if I didn't have all those other guys to contend with!" Like it or not, human beings spend a lot of time in small groups trying to please as many members as possible while simultanously trying to get something done. The decision or action may involve anything from picking a card game or a motion picture to electing a government official or passing a new law. For one reason or another it is very difficult to avoid such situations unless one avoids people altogether. Hence, it should come as no surprise to anyone to find that the topic of group decision making is virtually as old as Western philosophy itself. Indeed, perhaps the first great treatise on philosophy written in the West, Plato's *Republic*, deals precisely with the problems surrounding people's relations in groups. Plato is frying other big fish as well in that book, but good decision making for a group of people (the body politic) is at the center of his concerns.

One might ask (as my colleagues and friends often do ask): What does group decision making have to do with the philosophy of science? The short answer is: Plenty! Political science, sociology, economics, and psychology are all deeply concerned with decision making in groups, small and large. When a business student studies organizational behavior, he will study primarily decision making in organizations. Any science devoted to the study of human beings and their institutions must sooner or later pay attention to group activity, including group decision making. There really are precious few Robinson Crusoes to be studied. Accordingly, insofar as philosophers of science take seriously those features of the world that scientists take seriously, group decision making is an eminently suitable topic for the philosophy of science.

There is a less well known but extremely important by-product of investigations of group decision making. From a formal or strictly logical point of view, there is no difference between a group of people trying to make a decision and one person trying to make a decision involving a diverse set of considerations. For example, if a scientist sets out to weigh several different bits of evidence for and against his theory, he will find it necessary to proceed as if he had to reconcile

172

many conflicting views of the members of some group. He will have to insure fairness to all sides, determine how much weight to give to this or that side, when to call a draw and search for still new material, when to call a certain victory for one side, and so on. Formally or logically, such an individual is in the same boat as a group of individuals deciding to fish or cut bait. Thus, the study of group decision making also lays the groundwork for an investigation of problems related to the acceptance of scientific theories—a mainstream problem in the philosophy of science if there ever was one.

This paper has two aims: one fairly general, and the other specific. The general aim is to provide a basis for logical and empirical investigations of a particular type of group decision procedure. The type is distinguished by the inequalities in the "weight" of individual voters *and* their votes that are permitted. For reasons which will become clear shortly, if such procedures are practicable at all, they would seem to be so only for small groups, say, of two to fifteen people. Since most committees contain fewer than fifteen members, if such procedures are practicable at all, they should be practicable for committees. Hence, our discussion is oriented toward decision making in committees.

The more specific aim of this paper is to introduce and critically evaluate certain acceptable group decision procedures. We begin with a precise statement of the problem in section two. The third section introduces five informal, necessary conditions of adequacy for proposed solutions. Section four discusses the notion of "weights of influence" and defines seven fundamental types of distributions. This is followed by a presentation of six plausible "voting schemes" that involve the application of various "weights of influence." Sections six and seven examine the constructive work of sections four and five in the light of our adequacy conditions. The upshot of this examination is, unfortunately, that none of the six schemes appears extraordinarily attractive.

II. The Problem

Individuals are often called upon to share their decision making capacities with others: They are asked to serve on committees. Some committees exist to solve problems or make decisions. They frequently plan conventions, parties, and dances, i.e., they *choose* locations, entertainment, refreshments. They *select* textbooks, candidates for new positions, visiting professors, office furniture, etc.

To avoid difficult epistemological problems which need not be solved here, we shall assume that *a decision is made* if and only if a person or group come to accept a certain sentence which would normally be in the imperative or in the indicative mode. Since committees are social organizations with their own rules, norms, procedures, etc., ordinarily *a committee makes a decision* if and only if there is a sentence acceptable to the committee according to its operating rules. The operating rules of many committees are such that when there are conflicts of interest, the decision of the committee might not be similar to those of any of its members, e.g., a "compromise candidate" may become the first choice of a committee in which no member ranked him first.

Given the sense of "a committee decision," as explained in the last paragraph, the primary question we will attempt to answer in this paper is: How *should* the various decisions of the committee members be amalgamated so that the committee decision can be determined in the most acceptable fashion? Three points must be made immediately: First, although this question is normative, it cannot be answered solely by logical analysis and intuition. What we ought to do to reach more acceptable committee decisions depends, among other things, on the *composition* of and the *issues* considered by this or that committee: Proposals for optimal group decision making must be guided by some knowledge of the behavior, skills, resources, preferences, ambitions, etc. of human beings in fairly well-defined circumstances. Such knowledge is usually difficult to obtain. Secondly, in this paper the term "acceptable" will be used broadly to designate characteristics or dispositions frequently referred to by such (equally vague) terms as "competent," "wise," "good," "skillful," "enlightened," "rational," and so on. For the purposes at hand, this usage is innocuous and convenient. Finally, it should be emphasized that we are primarily concerned with the procedures, methods, or means used to arrive at committee decisions, *not* with the products, ends, or committee decisions themselves. Although in practice a peculiar committee decision often leads us to question the procedure used to obtain it (just as we might have suspicions about an argument schema which yielded a bizarre conclusion from ostensibly true premises), it is important to distinguish the former from the latter, the *decision* from the *decision procedure*.

III. Criteria of Adequacy

Lest we become dizzy chasing our own tails, something should be said about the *necessary* conditions of adequacy that might be applied to the amalgamation procedures. The following five are consistent (with one another) and seem plausible, although I suspect that they are neither independent nor in any technical sense complete. As we shall see later, they are considerably stronger than they appear.

174

(1) The procedures must be *free from coercion*: Committeemen must not be forced to vote contrary to their preferences. This guarantees each voter the opportunity to use his influence (the "weight" of his vote) exactly as he sees fit.

(2) The procedures must be self-consistent or *logically coherent*. This is merely a formal requirement to prevent contradictory decisions.

(3) The procedures must be *practicable* or manageable. Practicability is relative to people and issues; hence, only those procedures that are manageable for *most* (normal) people and *some* (realistic) issues will be accepted.

(4) The procedures must be *efficient*: They must take account of all of the relevant information available to a committee for a given decision at a given time. This requires not an everlasting search for relevant data, but the elimination of procedures that cannot process or use relevant data already possessed by committeemen.

(5) The *influence* of each committeeman should be *proportionate to* his relative competence: We suppose that a committee decision is to be reached by "combining" the decisions of all of its members. If the committee operates on democratic principles, each member's decision identically influences the final outcome (the committee decision). Each member receives one vote which, if cast at all, carries as much weight or has as much influence *formally, legally, arithmetically* on the final outcome as every other vote. (Informally, or in fact, of course, every group seems to have leaders and followers, both of which influence each other in more or less subtle ways.) The committee decision itself, then, is usually determined by the majority of the voters. According to (5), however, each member's vote should carry only as much weight or have as much influence on the committee decision as his competence merits. Moreover, since the composition of and the issues confronted by committees vary, (5) suggests that the status of a committeeman's competence should be regarded as *relative* to both of these variables.

In view of the vast amount of literature devoted to the problem of equality, my defense of (5) will probably seem much too brief and primitive. It is as follows: As the influence of a committeeman increases, the chances increase that committee decisions will be similar to his (provided, of course, that other things are equal and stable). As the competence (ability, rationality, etc) of a committeeman increases, the chances of his decisions being "correct" increase. Hence, as the influence and competence of a committeeman increase, the chances of committee decisions being "correct" increase. Similarly, it is easy to see that as the influence and incompetence of a committeeman increase, the chances of committee decisions being "*in*correct" increase. According to (5), acceptable procedures must contain some provision for the distribution of influence according to relative competence. Therefore,

175

insofar as this condition is met, the chances of committee decisions being "correct" will be increased and the chances of them being "incorrect" will be decreased. While this is not a guarantee that a committee will make "correct" decisions (or even *more* "correct" decisions *more* often), it does seem to be necessary for an acceptable group decision procedure.

Notice that the argument we have just presented does not commit us to the view that there is a "correct" decision which *is known* or even, practically speaking, *could be known* by certain committeemen. Of course, in certain fields (e.g., logic and mathematics) there are independent and explicit criteria of "correctness" for the solutions of most problems; so one frequently can and does *know* whether or not his decision is "correct." In many of the most interesting and important fields (e.g., human welfare, morality, law, religion, etc.), however, one seldom has such knowledge. But it does not follow that there are no "correct" or "incorrect" decisions for the various evaluative problems that arise in these fields.

IV. Weighting Voters

Even if one is persuaded by this argument for (5), which I am, it does not take us very far. For merely to say that we *ought* to distribute influence according to relative competence is not therefore to *know* how to do it. The remaining paragraphs of this section will be devoted to the problem of implementation.

Suppose we have a committee with *n* voting members:

$$M_1, M_2, \ldots, M_n.$$

Each member's vote will be assigned a *weight of influence* (or, for short, a *weight*) W which satisfies the following condition:

(1) The weight of every committeeman's vote will be a real number greater than or equal to zero, i.e.,

$$W_i \geq 0 \quad (i = 1, 2, \ldots, n)$$

In view of (5) in the previous section it is imperative that the weight of every member's vote is proportionate to his relative competence. But how are we going to measure relative competence? Clearly, the usefulness of our amalgamation proposals will be severely curtailed unless a fairly plausible answer to this question is produced. The solution offered here is not new and it is *at best* fairly plausible. Its greatest virtue is its apparent ease of application, i.e., it is relatively convenient.

176

Consider, for example, a committee with three members:

$$M_1 \quad M_2 \quad M_3.$$

The committee is supposed to select an appropriate logic text for a freshman course. The usual (democratic) procedure is to give each man one vote to cast as he sees fit. He may vote for this or that text or abstain altogether. Following this democratic tradition, each of our committeemen will also be given exactly one vote. Instead of asking each member to cast his vote on some question before the committee (e.g., the selection of a logic text), however, we will ask each to *distribute* the weight of his vote (which is unity) among each of the committeemen. The distribution of a given member's vote should indicate the relative competence of every voter in the judgment of that committeeman.

More precisely, we are asking *first* that each member *weakly order* every voter in accordance with his relative competence. That is, every committeeman is supposed to try to decide for all of the members whether

(i) M_i is more competent than M_k, or M_k is more competent than M_i, or M_i and M_k are equally competent;

and

(ii) if M_i is at least as competent as M_k and M_k is at least as competent as M_j, then M_i is at least as competent as M_j.

And *secondly*, we are asking that every member try to assign a number to every voter in accordance with his rank order to serve as an indicator of his relative competence. These numbers will be called "*initial* weights of influence" (or, "*initial* weights"). Moreover,

(2) Every initial weight will be a real number w in the closed interval from 0 to 1, i.e.,

$$0 \leq w \leq 1$$

and the sum of the assignments that may be given *by* every committeeman will be 1.

The following matrix illustrates a possible result:

	M_1	M_2	M_3	w_i
$\longrightarrow M_1$	1	0	0	1
M_2	$\frac{1}{3}$	$\frac{1}{3}$	$\frac{1}{3}$	1
M_3	$\frac{1}{2}$	$\frac{1}{2}$	0	1
W_i	$1\frac{5}{6}$	$\frac{5}{6}$	$\frac{1}{3}$	$3 = n$

Figure I

177

Here, reading the row following the arrow, M_1 has assigned himself an initial weight of 1 and left nothing for M_2 or M_3. We may assume then, that in the opinion of M_1 the other members of the committee do not know anything about logic texts. Or, to put the point another way, we may assume that M_1 believes the committee decision has a better chance of being "right" if it is similar to his own decision. M_2 believes his opinion is worth no more and no less than the others. So he distributes his initial weight equally among all the voters. M_3 figures that M_1 knows about as much as M_2 about logic texts and that he (M_3) does not know anything: So he distributes his initial weight equally between M_1 and M_2.

If a committee has n members then it has n votes to be distributed or, possibly, withheld. Then, by definition:

(3) The weight of every committeeman's vote equals the sum of the initial weights assigned to him by all members of the committee, i.e.,

$$W_j = \Sigma \ w_{ji} \quad (i \text{ and } j = 1, 2, \ldots, n)$$

for the jth committeeman according to each (ith) member.

In Figure I, the weight of M_1 equals the sum of the initial weights in the column below M_1, namely, $1\frac{5}{6}$. The weights of M_2 and M_3 are $\frac{5}{6}$ and $\frac{1}{3}$, respectively. The sum of these weights is $n = 3$, which happens to equal the maximum number possible for this group because no one withheld any of his initial weight.

Notice the efficiency of the voting procedure prescribed for these committeemen in comparison to the usual (democratic) procedure. Consider M_3's position, for example. Given the usual procedure, if M_1 and M_2 disagreed on a text then M_3 would have to *either* abstain altogether *or else* go along with M_1 or M_2. But in *both* cases he would be merely making the best of a bad situation. His "real" judgment is that M_1's views are about as reliable as M_2's. The voting procedure, however, forces him to act as if he had nothing at all to contribute (i.e., he abstains) or else to act as if he preferred the view of M_1 or M_2. In short, the usual procedure suppresses that tiny bit of competence that the committee recognizes in M_3. On the other hand, our procedure of weighting voters takes account of M_3's ability and allows him to make a contribution that reflects his own best judgment. It is *not* a judgment about the question at issue (i.e., about the best logic text), but about the relative ability of the other members of the committee to judge that issue. It is what many writers regard as a typically administrative decision—a decision about the ability of certain personnel to judge a certain issue, rather than a technical decision about the issue itself. The legitimacy and usefulness of administrative decisions can hardly be doubted, since it is difficult to imagine a highly complex organization such as an industrial corporation, university, or government agency without many people responsible for such decisions, viz., the administrators or managers. Therefore, if all other things are equal, then insofar as the procedure we are recommending provides a more efficient treatment of such decisions than the usual procedure, the former should be regarded as superior.

According to (1)–(3), the weight of each committeeman's vote must be a real number in the closed interval from 0 to n, i.e.,

$$0 \leq W \leq n \quad (j = 1, 2, \ldots, n).$$

If every committeeman distributes his initial weight in equal amounts among all the members of a committee, then each will receive an initial weight of $1/n$ from every member. Since there are n members, the result of such a distribution will be a *democratic weighting* with every committeeman's weight

$$W_j = 1 \quad (j = 1, 2, \ldots, n).$$

Formally, this result is indistinguishable from an *anarchic weighting* in which everyone keeps all of his initial weight. The different attitudes of democrats and anarchists toward legitimate government is reflected by the different paths leading to the equalitarian weights. The democrat assigns each committeeman the same weight and, therefore, is obliged to use numbers of supporters as a legitimizing criterion. The anarchist assigns only himself a weight and, therefore, is obliged to use his own preferences as a legitimizing criterion.

If every committeeman distributes his initial weight such that the weight of a single member, say, M_d is

$$W_d = n$$

the result will be a *delegatory weighting*. In effect the complete responsibility for the committee decision has been delegated to M_d. If the latter could be *born* into the position, there would be some justification for regarding this pattern as a *monarchic weighting*.

If the distribution of intitial weights is such that the total weight of fewer than half of the voters is greater than half of the total weight of the whole committee, the result is an *oligarchic weighting*. More precisely, if there are m committeemen such that

$$\Sigma \, W_i > n/2 \quad (i = 1, 2, \ldots, m)$$

although

$$m < n/2$$

179

the weighting is oligarchic. If the composition or membership of the set of m committeemen varies as the issues before the committee vary, it might be more appropriate to refer to the weighting as *polyarchic*. In practice, many organizations that operate *formally* with democratic weightings, operate *informally* with polyarchic or delegatory weightings.

Finally, if the distribution of initial weights is such that the weight of a single member, say, M_o is

$$W_o = 0$$

the result is a *disfranchising weighting*. Notice that no one can be disfranchised unless he chooses to be.

Obviously these weighting patterns are not mutually exclusive in pairs. While no pattern can be both oligarchic and democratic (or democratic and disfranchising), every delegated pattern must be oligarchic. Moreover, *a priori* there seems to be no good reason to suspect that the patterns defined here exhaust the interesting possibilities, i.e., these weighting patterns are probably not exhaustive.

More importantly, perhaps, it should be emphasized that it is *not* being claimed that the specification of the type of weighting pattern employed in a committee is *sufficient* to characterize the committee as, say, democratic. It is true, however, that the specification of the type of pattern is *necessary* for such a characterization.

V. Weighted Voting

Now that we have a procedure for weighting voters how should the weights be used to reach committee decisions? This can be neatly divided into the following two questions:

(A) How should weights be applied by voters?
(B) What proportion of the total weight (n) available should be required for a committee decision?

The latter question (B) has received considerable attention by legislators and philosophers, and we will not attempt to improve upon traditional discussions here. Instead, we will focus on the former (A) and try to present some options that have received little or no attention. In particular, we will consider six voting schemes. For each of the schemes, it will be assumed that the decision (item, solution, policy, candidate, etc.) which receives the greatest support or has the largest weight will be accepted as the committee decision. Roughly speaking, this assumption is equivalent to answering (B) with: A plurality or simple majority is required for a committee decision. This is not an especially profound or unproblematic answer, but it will be adequate for our purposes. Above all, it should be remembered that (A) and (B) are quite different questions, and that success or failure with either does not imply success or failure with the other.

Question (A) is vague. It might be answered by such disparate remarks as: in accordance with their consciences; with caution; happily; all together; in bits and pieces, etc. The last two replies suggest the sense in which we are interested. To begin with, we might apply weights as follows:

> *Total weight scheme:* Each committeeman puts the total weight of his vote on a single decision (candidate, option, etc.), or he withholds all of it.

For example, suppose the weights assigned to a five-membered committee are

	M_1	M_2	M_3	M_4	M_5
W_i	1	.5	1.3	.7	1.5

Figure II

and there are three alternative decisions to consider

$$D_1 \quad D_2 \quad D_3.$$

Then on the total weight scheme, the result of a vote might be

	D_1	D_2	D_3
M_1	0	1	0
M_2	.5	0	0
M_3	1.3	0	0
M_4	0	0	.7
M_5	0	1.5	0
Total	1.8	2.5	.7

Figure III

That is, M_1 puts all of his weight behind D_2; M_2 puts all of his weight behind D_1, and so on. The result is that D_2 becomes the committee decision because it has the most support.

The total weight scheme is perfectly straightforward. Each man either applies all of his influence (i.e., the total weight of his vote) to a single decision or he withholds all of it. Sometimes a committeeman regards certain alternative decisions as equally acceptable, however. Or, faced with three or more options, he is frequently able to order them weakly. If *any* of the members of a committee can so order the various alternatives before them, then it would be useful (i.e., more

181

efficient) to give them the option of dividing the weight of their votes according to their preferences. This involves two closely related assumptions. First, it must be granted that weights of influence, which are already indicators of relative competence, may also be used as indicators of preferences. And secondly, it must be granted that people's preferences are comparable, e.g., if M_1 and M_2 rank D_1 above D_2 and the latter above D_3, we would *grant that* they both "feel about the same" about the three decisions *instead of* insisting that, say, M_1 prefers D_1 to D_2 much more than M_2 prefers D_1 to D_2, and so on. While the dual role of weights of influence seems harmless enough, we will have more to say about the weak ordering and interpersonal comparisons of "utility" in section seven.

Given the above assumptions, we might employ a

Split weight option scheme: Each committeman distributes the weight of his vote among the decisions in accordance with his preferences, or he withholds any part or all of it.

For example, consider the five-member committee above, faced with the same three decisions. A possible result of the split weight option scheme for the voters in Figure II might be

	D_1	D_2	D_3	W_i
M_1	.5	.5	0	1.
M_2	.5	0	0	.5
M_3	.8	.5	0	1.3
M_4	0	.2	.5	.7
M_5	.5	1	0	1.5
Total	2.3	2.2	.5	$5 = n$

Figure IV

Here M_1 distributes the weight of his vote between two equally attractive (acceptable, preferred, etc.) decisions D_1 and D_2; M_2 keeps his vote intact, and so on. The result is that D_1 narrowly becomes the committee decision.

A major advantage of the split weight option scheme over the total weight scheme is that the former *takes account* of discernable preferential differences which the latter cannot. Or, to put this important point in a slightly different way, a committee decision based on the split weight option scheme is a function of more subtle or precise judgments than a committee decision based on the total weight scheme. Hence, the former scheme is more efficient than the latter.

Instead of allowing committeemen to divide their weights to indicate their preferences, we might provide each voter with an additional unit to distribute. More precisely, we might let each voter assign every decision (item, policy, etc.) a real number U in the closed interval from 0 to 1 called a "utility," with the sum of utilities assigned by each member equal to or less than one. Here the term "utility" means nothing more than the value (desirability, satisfactoriness, etc.) of a decision according to a committeeman. An example of the result of such assignments might be

	D_1	D_2	D_3	sum
M_1	.5	.5	0	1
M_2	1	0	0	1
M_3	.7	.3	0	1
M_4	0	.3	.7	1
M_5	.4	.5	.1	1

Figure V

If U_{ij} is the utility of the ith decision according to the jth member then the total *weighted-utility* of that decision for that member is

$$U_{ij} \ W_j.$$

Now we may define the

Total weighted-utility scheme: Each committeeman assigns every decision a total weighted-utility (i.e., the product of the total weight of influence of his vote and the utility he assigns to each decision).

For example, given the utility assignments in Figure V and the weights of influence in Figure II, we have

	D_1		D_2		D_3	
M_1	(1)	(.5)	(1)	(.5)	(1)	(0)
M_2	(.5)	(1)	(.5)	(0)	(.5)	(0)
M_3	(1.3)	(.7)	(1.3)	(.3)	(1.3)	(0)
M_4	(.7)	(0)	(.7)	(.3)	(.7)	(.7)
M_5	(1.5)	(.4)	(1.5)	(.5)	(1.5)	(.1)
Total weighted-utility sum	2.51		1.85		.64	

Figure VI

The result of applying the total weighted-utility scheme with the given numerical assignments is a clear victory for D_1.

In order to obtain greater uniformity in our utility assignments, we might employ the sort of marking system introduced by Jean-Charles Borda: We would assign 0 utility to the least preferred

183

alternative, 1 to the alternative immediately above that, 2 to the third highest, and so on until we reach the end of the lot. Equally preferred alternatives are assigned the same number. For convenience, we will call such indicators "Borda-utilities." The preferences ranked in Figure V would be indicated as follows in terms of Borda-utilities.

	D_1	D_2	D_3
M_1	1	1	0
M_2	1	0	0
M_3	2	1	0
M_4	0	1	2
M_5	1	2	0

Figure VII

Notice that while alternatives with Borda-utilities of n have exactly n *ranks* below them, they may have more or less than n *alternatives* below them. For example, M_2 distinguishes two preference levels. So the top level has a Borda-utility of $n = 1$, and there is one level but two alternatives below D_1.

Now we may define a

Weighted Borda-utility scheme: Each committeeman assigns every decision a weighted Borda-utility (i.e., the product of the total weight of influence of his vote and the Borda-utility he assigns to each decision).

For example, given the Borda-utility assignments of Figure VII and the weights of influence of Figure II, we have

	D_1		D_2		D_3	
M_1	(1)	(1)	(1)	(1)	(1)	(0)
M_2	(.5)	(1)	(.5)	(0)	(.5)	(0)
M_3	(1.3)	(2)	(1.3)	(1)	(1.3)	(0)
M_4	(.7)	(0)	(.7)	(1)	(.7)	(2)
M_5	(1.5)	(1)	(1.5)	(2)	(1.5)	(0)
weighted Borda-utility sum	5.6		6		1.4	

Figure VIII

Hence, the result of applying the weighted Borda-utility scheme with the given numerical assignments is a victory for D_2.

If we provide additional units such as utilities and Borda-utilities to be distributed *and* allow committeemen to split the weight of their votes, then we may define a

184

Split weight utility scheme: Each committeeman assigns every decision a split weight utility (i.e., the product of some or all of the weight of influence of his vote and the utility he assigns to each decision);

and a

Split weight Borda-utility scheme: Each committeeman assigns every decision a split weight Borda-utility (i.e., the product of some or all of the weight of influence of his vote and the Borda-utility he assigns to each decision).

Using the Borda-utility assignments of Figure VII and the weights of influence of Figure II, Figure IX illustrates a possible result of applying the latter scheme.

	D_1		D_2		D_3	
M_1	(1)	(1)	(.5)	(1)	(.5)	(0)
M_2	(.5)	(1)	(.2)	(0)	(.5)	(0)
M_3	(1.3)	(2)	(.3)	(1)	(.8)	(0)
M_4	(.7)	(0)	(.7)	(1)	(.7)	(2)
M_5	(1.5)	(1)	(1)	(2)	(.5)	(7)
split weight Borda-utility sum	5.6		3.5		1.4	

Figure IX

Hence, the result of applying this scheme with the given numerical assignments is a victory for D_2. (Notice that every committeeman is permitted to multiply the Borda-utility he assigns *to each decision* by a number no greater than his total weight.)

By a line of argumentation analogous to that used to show that the split weight option and total weight-utility schemes are equivalent, it may be shown that the split weight utility and split weight Borda-utility schemes are both equivalent to the former two schemes. Thus, in view of the unnecessary labor involved in the additional computations required for the application of the latter three in comparison with the split weight option scheme, there is no doubt that the other three schemes will never be practicable.

The following matrix summarizes the six schemes we have considered. The total weight scheme comes about by answering question (A) with "as a whole" and not providing any additional units to be distributed. If (A) is answered by "as a whole or in parts" and no additional units are provided, the result is the split weight option scheme. The remaining entries in the matrix arise analogously.

185

	No additional units provided	Additional units provided
Use all or no weight	total weight	total weighted-utility weighted Borda-utility
Use some, all, or no weight	split weight option	split weight utility split weight Borda-utility

VI. Objections and Replies: Weighting Voters

The constructive work of this essay has now been completed, and it is time to appraise our results according to the adequacy criteria introduced in section three. Six solutions to our basic problem (viz., how should the decisions—with respect to some issue—of the members of a committee be amalgamated in order for the committee decision to be determined in the most acceptable fashion?) were suggested in section five. Each of these six employed the procedure for weighting voters that was described in section four. So, if there are no acceptable procedures for weighting voters, then the practical value of the six schemes of weighted voting must be nil. It will be convenient then, to consider objections to the procedure for weighting voters prior to and in isolation from objections to particular schemes of weighted voting.

How does our procedure for weighting voters fare with respect to our five criteria in section three, viz., (1) freedom from coercion, (2) internal consistency, (3) practicability, (4) efficiency, and (5) influence proportionate to relative competence? Taking the least problematic first, our procedure seems (to me) to satisfy the conditions of freedom from coercion, internal consistency, and efficiency. But the other two requirements require some discussion.

To begin with, it might be objected that our procedure violates (5) because it begins with an equalitarian distribution of influence in the form of a single vote worth exactly one unit. That is, *before* anyone's competence is estimated, every committeeman is given a vote whose initial weight is equal to that of every other. But according to (5), such a distribution is permissible only if the competence of every committeeman is equal to that of every other. At this first stage in our procedure, however, we do not know how competent any committeeman is. Hence, we cannot justify an opening equalitarian distribution. Moreover, if we *could* justify such a distribution, then the whole procedure of weighting voters would be a waste of time: All would always receive the same weight. But the first step in our procedure must be either unjustified or justified. Therefore, it must either violate (5) or trivialize the whole procedure.

The horns of this dilemma are more apparent than real. The first is problematic, but the second is based on a misunderstanding. Let us consider it first. Recall that in our procedure each committeeman makes two different *types* of decisions, one on the abilities of the other members and the other on the issue(s) before the committee. Our opening equalitarian distribution only presupposes that every committeeman is equally qualified to make the first type of judgment. If *this* assumption is warranted, then every member's *opinion about every committeeman's judgment* about the issue(s) before the committee is equal to every other member's opinion. It does *not* follow that every member's *opinion about the issue(s)* before the committee is equal to every other member's opinion. Hence, the whole weighting procedure cannot be trivialized by an opening equalitarian distribution.

The first horn of the dilemma cannot be dispatched so swiftly. Indeed, we must grant immediately that we have no reason to suppose that this ability is equally distributed among all or most people. What we would like to substitute for an opening equalitarian distribution is a distribution based on some reliable independent test. But how to select such a test? And this question seems to lead us straight to section two of this paper, *or* to an infinite regress of reliable independent tests, *or* to some more or less arbitrarily final reliable independent test. No doubt, most will prefer the last alternative and haggle over the cut-off point. The opening equalitarian distribution suggested in our procedure is merely a convenient (admittedly early) cut-off point which should be adequate for most purposes. After all, even the usual democratic procedure (which is less efficient than any of ours) has been regarded as adequate for most purposes. If the *cost* of an erroneous committee decision is very high, we would expect much more elaborate precautions to be taken to increase the chances of arriving at a more acceptable opening distribution.

Briefly, then, my reply to the first horn of the dilemma is as follows: The objection is sound but not totally destructive. It reveals a defect in our procedure which becomes more or less serious as the cost of errors increases or decreases, respectively, and which further research should attempt to eliminate or reduce as much as possible.

Supposing we had solutions to these problems, would our procedure then be practicable—condition (3)? It seems that our procedure may be impracticable for *at least* eight reasons:

1. Committeemen may find the task of weakly ordering their colleagues in accordance with relative competence too *complicated.*

187

Competence is very likely multi-dimensional, and the dimensions may be extremely difficult if not impossible to compare. For example, a certain committeeman may be very good in identifying, say, the pedagogical strengths and weaknesses of a textbook, but very bad at recognizing factual or formal errors. Should such a committeeman be assigned a high or a low weight? Should a committeeman who is good at *both* tasks be assigned a weight that is *twice* as great as that of a committeeman who is only good at one? Are the two skills "really" equally important? Even if these problems could be solved, the cost of solving them may be too high *given* the benefits obtainable from our procedure *and* the availability of others. It is not merely this or that complication in the abstract that may be objectionable, but that in comparison with other procedures, the cost of untangling the complications in ours may be too high and the likely rewards too low to vindicate its application.

2. The determination of relative competence may produce *disputes*, *factions*, and *enemies* that might otherwise be avoided. For example, a member may not appreciate receiving a low weight from a voter to whom the former has just assigned a low weight. Since committee meetings will probably take more time with our procedure, we may expect greater fatigue, irritability, impatience, and ennui: Even trivial issues may then become highly problematic.

3. Our procedure may be regarded as too *"brutal"* because it does not provide any face-saving safeguards for perennial "light-weights." Even if the weights are never made public (which merely requires the help of a discrete and honest assistant, or a machine), it is bound to occur to some people that in their *own* opinion the judgment of certain committeemen is almost always nearly worthless. Worse, if the split weight option scheme is used and everyone must be informed of their colleagues' views about their relative competence, then perennial "lightweights" are going to be repeatedly embarrassed and probably embittered.

4. Our procedure may breed *irresponsibility* in those who, for a certain issue, have been assigned a fairly low weight. The low weight assigned to a committeeman may have the effect of a self-fulfilling prediction: Everyone expects him to contribute very little, so he expends very little effort and, consequently, he has very little to contribute.

5. Committeemen may nullify the opportunity to determine the relative competence of their collegaues by employing a *minimax loss* strategy. According to this strategy, a voter should distribute his initial weight to insure himself the smallest possible loss. His options are to give away none of it, give away part of it, or give away all of it. Clearly,

he can protect himself against any loss at all by simply keeping his vote all to himself. If everyone employs this strategy, however, then the result is an anarchic weighting. And if enough people employ this strategy often enough, our procedure would have to be regarded as redundant.

6. It is likely that committeemen will *misjudge* the relative competence of their colleagues. They will commit "errors of leniency" (i.e., assign more weight to those they like), "central tendency" (i.e., avoid assigning extreme weights of 0 or 1), "logic" (i.e., assign similar weights for skills that seem to be "logically related"), "contrast" (i.e., assign weights roughly inversely proportional to those they assign themselves), and "proximity" (i.e., assign similar weights for skills that are considered at roughly the same time or in the same context). They will fall prey to the "halo effect" (i.e., assign weights in accordance with their "general impression" of the individual rated) and the "Matthew effect" (i.e., assign higher weights to the "biggest names" regardless of their particular competence in a given situation). They will be influenced by "prestige considerations," and, as the ambiguity of the situation increases, by social pressure to "conform." Hence, instead of indicating relative competence, our weights will be meaningless numbers misleadingly suggesting precision and accuracy.

7. There is some evidence that less competent people are less influential than more competent people anyhow. An attempt to *formally* weaken the former and strengthen the latter is, thus, unnecessary.

8. If a committeeman believes that some other member of a committee is especially competent, then, rather than giving the latter some of his weight (certainly losing formal influence and flexibility and possibly informal influence), he would be wise to keep all of his weight and simply allow himself to be guided by the other member's decision. After all, from the former's point of view, the final result of giving the latter his weight to put behind a certain decision is no different from putting it in the same place under the other member's supervision. Hence, because committeemen have something to lose by distributing their weight and nothing to lose by keeping it, few or no distributions should ever occur.

Let us consider each of these objections in turn:

1. It must be granted that *some* people may find the task of weakly ordering their colleagues in accordance with relative competence virtually impossible. Our procedure does not, therefore, *require* such an ordering from anyone. The ordering is *requested*, and if most people cannot satisfy the request most of the time, then our procedure cannot be expected to accomplish anything most of the time. Neverthe-

less, our procedure provides an opportunity to use whatever resources happen to be available, which seems to make it superior to procedures without this provision. Moreover, and fortunately, there is no evidence of a widespread human inability to perform such tasks. On the contrary, while it is difficult to appraise the relative competence of this or that scholar in a certain area, such appraisals are very common: Thousands of employers are required to do just that, and they are frequently promoted for doing it well and discharged for doing it poorly. Hence, the real question seems to be: Are the rewards of untangling the complications in our procedure high enough to justify the costs? And in the absence of a definite issue, committee, and alternative procedure, this question seems to be unanswerable.

2. There seem to be good reasons for believing that our procedure will not create an inordinate number of disputes. First, people frequently judge the merits of candidates for various positions of leadership, and our procedure is merely an extension of this common practice. Secondly, certain precautions may be taken to reduce the chances of disputes. For instance, one should avoid using our procedure in situations in which the likelihood of misunderstanding is known to be high, e.g., in meetings involving people with quite specific and, perhaps, antagonistic *roles* such as labor and management, or officers and enlisted men. Perhaps committeemen might be impressed with the idea that they are all part of a single "team" and that the less energy they expend attacking each other, the more they can expend on the issue before the committee. Again, the reward and punishment structure might be designed to militate against such phenomena, i.e., let the "payoffs" go to those who do not get involved in disputes and factions, and the penalties go to those who do.

3. The third objection seems to suggest a nice problem of balancing the demands of morality against those of rationality. In fact, the problem is to balance the demands of one moral prescription against those of another. On the one hand, committeemen are morally obliged to try to avoid embarrassing their colleagues. On the other, they are morally obliged to try to make as "correct" a decision as possible. Hence, they must weigh the moral cost of embarrassing (and, perhaps, embittering) some members against that of making a less "correct" decision. If it costs more (i.e., seems to create more feelings of guilt) to identify and isolate the "light-weights" than it does to accept a slightly "incorrect" committee decision, then a committeeman ought to accept the latter; otherwise, he ought to choose the former, unless the costs are equally balanced. In the latter case he might flip a coin.

It would be a mistake to assume that "light-weights" *must* be embarrassed or embittered: There is some evidence that the reaction of individuals to participation in or exclusion from decision making that affects their lives is significantly influenced by their *expectations*. Someone who expects "the other fellow" to be responsible for this or that may be perfectly happy to be assigned a lighter weight. Moreover, it should be noted again that if the variety of issues considered by a committee is large, the probability that a certain committeeman will always be judged a "light-weight" is small. Most people tend to be specialists by inclination and aptitude. Hence, if the issues before a committee *vary*, it is likely that no one will be a perennial "light-weight" *or* "heavy-weight," i.e., the *typical* weighting will probably be polyarchic.

4. It seems that we might suggest with equal plausibility that those committeemen from whom everyone expects a great deal will be encouraged to work harder and, consequently, will be able to contribute more than anyone expects. But this reply merely reproduces the same unwarranted assumption on which the objection is based. That is, there is no reason to assume that committeemen must be weighted long before the final vote on the issue is taken. Indeed, there is no doubt at all that the weighting should immediately precede the voting and that *both* should take place after a discussion of the issues. Moreover, every committeeman should be given an agenda far enough in advance of the meeting for him to become familiar with the issues. Given these tactics and our procedure allowing committeemen to gain complete control of a committee decision, it would seem to be difficult for a person with average ability to justify an attitude of impotence.

5. The application of a minimax loss strategy is not incompatible with assigning initial weights according to judgments of relative competence, i.e., it does not entail an anarchic weighting. If someone regards a less "correct" committee decision as a greater loss than a certain amount of personal influence on that decision, then the minimax strategy would lead him to distribute his initial weight in support of competence, rather than merely in support of himself. In effect then, he would be applying the strategy to the following alternatives:

1. Keep everything and obtain a less "correct" committee decision.
2. Distribute part and obtain a more "correct" committee decision.

The second alternative might well represent the smallest possible loss. The important unanswered question is: Will enough people value "correctness" more than personal influence to make our procedure effective?

6. The evidence for these suggestions is fairly strong. We can only try to devise ways to minimize the effects of such propensities. One way

to tackle the problem is to construct a set of minimum standards according to which initial weights must be distributed. For example, in the case of the selection of a textbook, we might give so many points for having taken so many courses, for having taught so many, for having published so many related papers, etc. Other more reliable tests would have to be developed for other types of issues. Alternatively, the committeemen could present their "credentials" with precise and standardized descriptions in order to reduce vagueness and ambiguity and the temptation to interpret perceptions in a more or less arbitrary fashion.

7. Insofar as events are bound to take place exactly as we prefer, with or without our (formal) efforts, such efforts must be regarded as unnecessary. But the ultimate triumph of wisdom over ignorance is hardly inevitable. Hence, we seem to be obliged to take *some* steps to "further the cause of rationality and good decision making." The formalization of influence in some situations might be *a* step in the right direction.

8. It seems clear that it would sometimes be wiser to *copy* a more competent member's voting behavior than to give him all or part of one's weight. In other situations, however, the latter strategy seems preferable. For example, suppose that one is not only interested in the "correctness" of a committee decision, but also in appropriately allocating the rewards or punishments resulting from that decision. The distribution of weights could be used as a basis for fixing responsibility and for fairly allocating rewards and punishments. Similarly, weight distribution provides an explicit record of committeemen's views of one another's competence which could be useful for choosing operating procedures and personnel. Again, it is often important to know whether a committee decision has received a certain amount of support on its own as it were *or* simply on the strength of its supporters. After all, to return to a point raised in section four, it is one thing to endorse a particular alternative and quite another to endorse the endorser. A ten-membered committee decision with one knowledgeable supporter and nine administratively competent, but technically ignorant supporters seems to have less support in some sense than the same decision with nine knowledgeable supporters and one supporting administrator. While weight distribution (according to the procedure suggested here) might not alter the final number of points received for such a decision in either case, it would provide some intuitively useful extra bits of information whose value would tend to vary directly with the error and implementation costs of any decision. Of course, one could probably gather the extra information by additional research of some sort, but weight distribution always

provides such information and always makes the subtle difference in the *kind* of support received by any alternative completely explicit.

VII. Objections and Replies: Weighted Voting

After reviewing the objections and replies to the procedure of weighting voters, we have not yet decisively made a case for or against it. There are still a number of loose ends, some of which require empirical investigation more than logical analysis. In this section we shall assume, however, that our procedure for weighting voters is more or less acceptable and ask: How do the various schemes of weighted voting fare in the light of our five adequacy criteria?

Taking the least problematic point first, I think that we may grant the internal consistency—condition (2)—of all six schemes. It is certainly true that if one requires some sort of piecemeal comparison of alternatives (e.g., pairwise comparison, triplewise, etc.) *and* assumes that both individual and group systems of preferences must be transitive, then all of these schemes are liable to generate the so-called "paradox of majority voting" in certain situations. But it seems to me that there is no good reason to expect group systems of preferences to be transitive or to require piecemeal comparisons of alternatives. The "paradox" is something of a red herring.

The *paradox of majority voting* may be explained as follows: Suppose you have a three-membered committee with members

$$M_1 \; M_2 \; M_3$$

charged with the election of one of three candidates

$$D_1 \; D_2 \; D_3$$

It is assumed that:

(1) Majority rules.
(2) Each member's set of preferences is weakly ordered (as explained in section four above).
(3) The committee's collective preferences are also weakly ordered.
(4) Candidates are considered two at a time, i.e., a committeeman looks at D_1 and D_2 without considering D_3, then D_1 and D_3 alone, etc.

In fact, the three committeemen end up rank ordering the candidates thus:

M_1	M_2	M_3
D_1	D_2	D_3
D_2	D_3	D_1
D_3	D_1	D_2

Examining the rank orders, we find that two people prefer D_1 over D_2, namely, M_1 and M_3. Since two people constitute a majority in a three-membered committee, we must say that the committee collectively prefers D_1 to D_2. That's what majority rule—assumption (1)—means in this case.

Similarly, however, two people prefer D_2 to D_3. So, the committee prefers D_2 to D_3. But according to our third assumption, committee preferences are transitive. That just means that if the committee collectively prefers D_1 to D_2 and D_2 to D_3, then it prefers D_1 to D_3. Fair enough.

The trouble is, if you count the number of members who have an opposite preference, namely, D_3 over D_1, you find a majority! That is, M_2 and M_3 prefer D_3 to D_1. Hence, we have the paradox that the committee collectively prefers D_1 to D_3 *and* D_3 to D_1.

As I said above, there seems to be no good reason to assume (3) or (4) above—(2) either, for that matter; hence I am not troubled by the paradox.

It may be recalled that in section five we showed that the total weight scheme is inefficient—violating (4)—in the sense that it is liable to be unable to take account of available information. Similarly, the weighted Borda-utility scheme must be inefficient because Borda-utilities vary at regular intervals whether or not the preferences of committeemen vary that way. For example, if someone ranks D_1 above D_2 and the latter above D_3, then the Borda-utility of D_1 is twice that of D_3 whether or not the committeemen believe D_1 is twice as "good" as D_3. While we are not *bound* to use the whole numbers 0, 1, 2, 3 . . . as Borda-utilities, we are bound to use *some* regularly increasing set (e.g., 0, 1/3, 2/3, 2/3 . . . ; 0, 1/4, 1/2, 3/4 . . .) because it is this very regularity that distinguishes this scheme from the total weighted-utility scheme. Hence, whenever committeemen have preferences that do not vary at regular intervals, they will be unable to "feed" this information into the weighted Borda-utility scheme. Hence, it too violates (4). Since the split weight Borda-utility scheme provides some means of breaking the rigid Borda-utility mould (namely, by appropriately splitting one's weight), it does not violate (4). Furthermore, for roughly the same reason, neither do the other three schemes.

The total weight scheme seems to be free of any peculiar problems of practicability—condition (3). As we saw in section five, however, the total weighted-utility, split weight utility and split weight Borda-utility schemes are all impracticable because the labor costs of applying them are greater than the costs of applying the equivalent split weight option scheme. Furthermore, with the exception of the total weight scheme, they all presuppose (i) the ability of some committeemen to weakly order their preferences for various alternatives and (ii) the interpersonal comparison of utility. Both of these assumptions have been debated so exhaustively in the literature that it seems very unlikely that anything novel can be said here. So, I will restrict my "defence" to the following brief remarks.

In defence of (i) it should be noted that people are frequently (though by no means always) able to order their preferences weakly, and that this must be regarded as favorable evidence for the view that, as far as this problem is concerned, our voting schemes will frequently be practicable. In defence of (ii) it must be admitted that the very existence of fairly stable wage and price systems is evidence that some services and commodities must have roughly the same value for many people. That is, if people of roughly equal means are willing to pay the same price for a certain commodity or do the same work for the same salary, then it seems more likely that they are receiving the same rather than different amounts of satisfaction from the exchange. Hence, this second problem seems to be primarily one of constructing appropriate technical "devices" to obtain reliable measures of comparability. Still, it would be gratuitous to assume that these schemes did not face a serious problem of practicability.

IX. Conclusion

Insofar as we have succeeded in providing a more or less systematic basis for logical and empirical investigations of group decision procedures which permit formal or explicit inequalities among voters and votes, our more general task has been completed. Insofar as we have been able to throw some light on the relationships, strengths, and weaknesses of various more or less plausible solutions to the problem of selecting an acceptable group decision procedure, our more specific task has been accomplished. Perhaps the most appropriate summary of our results on the latter score would be a loose paraphrase of a remark made by Sir Winston Churchill: The usual (democratic) procedure is a very bad form of government, but it is every bit as good as all the others.

Technology in Perspective

JOHN W. ABRAMS

There are few persons who have not at some time used the word, technology. They would no doubt agree that it had some connection with the work of a medieval smith at his forge, a puffing steam engine, a noise-filled factory with moving belts driving machines, a television set, or a nuclear power plant. The common element in the above is their origin in an activity we may call technology. Can we say more about this activity? Is its essential character a description for certain artifacts, or is it, as many say, a determining factor in society? Is there a philosophy of technology apart from the philosophy of science? Are they similar? These questions should be explored, even if they may not permit of satisfactory answers.

The question "What is science?" has been the subject of much study. The question "What is technology?" may be today more pressing, yet it has received less serious consideration. The common agreement that technology has been, is, and will continue to be a determining factor in the future of life on this planet conceals the measure of disagreement as to what is to be considered as technology. Examination of the burgeoning current literature on philosophy of technology reveals sufficient differences in delineation of the field that it is possible to arrive with logical consistency at positions and interpretations which are diametrically opposed, and which are incompatible in terminology as well as conclusions.

It is possible to trace much of the contradiction to the definitions of technology used in the various studies. In fact, many of the studies themselves are essentially attempts to establish a definition which can then be used as a basis for analysis and argument. If success in these attempts is to be measured by the degree of establishment of a group consensus, this goal is far from having been achieved. The failure is not in the definition alone; it often lies in the philosophic presuppositions of the author.

The philosophy of technology has been implicitly or explicitly the subject of several symposia in the last decade. Their convocation indicates sufficient agreement to arrange a programme and sufficient disagreement that there is no problem finding speakers with different views. As the question "What is technology?" is not resolved, the difficulty of analysing an undelineated area of activity and knowledge continues to keep us in confusion and disagreement.

Whereas many so-called "core" definitions have been explicitly given, they are almost invariably recognized as inadequate by the authors themselves, who proceed to expand, criticize, and interpret their original succinct statements. Eventually they end up with discussions and conclusions on the nature of technology, its relation to science, society, and philosophy, or its significance to the past, present, and future of man.

There is also a considerable body of literature in which an explicit definition is not given—even to be subsequently used as a whipping boy—and the author proceeds, often successfully, to adumbrate his thesis that technology is important and must be embraced, or that it must be rejected or controlled if mankind is to survive. This group of authors cannot be dismissed. They are influential despite or maybe because their incomplete definition permits the reader personally to fill in the gaps and thus to meld his own interpretation with that of the author. These authors are mainly scientists, but those who also write on the meaning or influence of technology come from many fields of specialization. By calling on their own special expertise they have brought new points of view and unearthed new aspects worthy of consideration. We have had philosophers of science, historians of science, social historians, economic historians, mathematicians, sociologists, political scientists and economists, futurists, generalists, engineers, theologians, and even "traditional" philosophers considering various aspects and influences of technology. The number interested endorses the recognition that there is an area of importance; the variety of their approaches and the different aspects signaled out for special attention indicates the breadth of concern.

A manifestation of the undefined boundaries of technology is a propensity to coin new words or to make specialized use of old ones. We have technics (Mumford), technique (Ellul), media (Innis and McLuhan), as well as numerous variants with the prefixed descriptive adjective technical or technological. Each represents what McLuhan calls a "probe"—an attempt to bring forward for consideration an aspect of a not-necessarily-defined area with the object of fostering discussion and perhaps action.

The variety of meanings, even when they are clear, militates against a traditional approach toward a core definition which would consider the intersection or overlap of a number of proposed definitions as a nuclear starting point. The intersection is not barren; but it is innocuous if not vague. We shall propose here a new, but admittedly inadequate, definition, introduce specific meanings for old terms (tool, technique, etc.), and compare this framework with selected treatments. Such comparison is not new; our innovation lies partly in the method of comparison and partly in an open acknowledgement that the treatment is tentative and merely an early stage in what will, one hopes, be an iterative procedure. This first iteration is not expected to do more than lead to a second which may be more satisfying.

Whether this procedure will converge to a limit is not in question. It will not affect our method or results. At this stage we shall not reach the limit, if it exists. There may well be such a limit—a Platonic Idea of Technology. We shall, however, with Plato acknowledge the difficulty if not the impossibility of attaining knowledge of this Idea. Technology as we shall treat it has a closer relationship (or a more attainable one) with Becoming.

Attention should be directed to the number of studies of technology which use an historical approach. Despite neglect of the history of technology until the last few decades, historians have always recognized the importance of technological input and technological change in directing the path of history, although they have avoided detailed analysis. The appellations, Stone Age, Bronze Age, etc. bear witness. There has also been a recognized connection between the content of technology and the modern idea of progress since the inception of the latter around the start of the eighteenth century (e.g., Francis Bacon and *The New Atlantis*). More recently, attention has been directed to the apparent relationship between the scope of what has at any point in time been considered technology, and the contemporary natural and social philosophy. Successive definitions are found either to extend or restrict the scope. For example, as scientists have changed their point of view from a mechanistic, Newtonian world picture to one based in field theory, the scope of what has been called technology has been extended by many to include more than the strictly mechanistic as had been the earlier view. If, however, we trace the word technology to its Greek roots, we find that in antiquity it certainly transcended the mechanical. We shall reflect such changes in our definition. The incompatability of many studies can be traced to the implicit adoption of ideas, already obsolescent. Nineteenth-century definitions are not appropriate to the twentieth century. The historical studies give us perspective to avoid errors of this type.

The relationship between science and technology is of prime importance today. We must recall that science has also changed both in content and structure. Although there are many active issues in the philosophy of science, the subject is better explored than the philosophy of technology. We must consider whether technology is merely applied science as is frequently implied, if not claimed outright; whether they are two distinct areas with a possible intersection or overlap; or even if science is a sub-set of technology. To do this we shall have to propose our own definition of science, a formidable task.

The most frequent distinction made between science and technology lies in the goals presumed for those engaging in them as activities: The scientist is supposed to seek pure knowledge of a particular sort, the technologist a specified, particular defined goal. As the knowledge sought by the scientist may aid in achieving the goal sought by the technologist, the two activities may be undertaken by the same person at the same or different times. With the above as the sole distinction, it would generally be impossible to distinguish between the actions without access to the minds of the actors. As such access is normally impossible, we shall not puruse this line of possible differentiation.

Science and technology will each be considered here to comprise both a body of knowledge and an activity. The activity consists in both cases of extending the structure and content of a body of relevant knowledge. Discussion of a further activity, application, will be deferred. For convenience the terms science and technology are used as if they possess an internal autonomy—a virtual anthropomorphization. We shall adhere to this convenient and common practice, but recognize it as no more than a convention. Even if we (or others) say "science shows" or "technology leads to" we shall mean that human initiation of a procedure within the accepted canons of either science or technology has resulted in certain conclusions or demonstrations. We shall not consider whether either has a unique epistemology.

We take the activity which we call science to be the endeavour to include all natural phenomena within a pattern. The goal of the pattern is to embrace all phenomena observed by our senses directly as well as those observed with the aid of instrumentation both within and without a contrived experimental framework. The goal has not been reached; the pattern is incomplete. Meanwhile, we have established canons of acceptability—in fact, conventions—which we apply as criteria for inclusion of new material or concepts within the accepted pattern and rejection or modification of the old. These canons mainly consist of a series of logical axioms and a rule of experimental reproducibility. The canons are not immutable: They have undergone changes—principally when a so-called scientific revolution or paradigmatic shift (Kuhn) has taken place. Even in these times of scientific change, however, most of the criteria of acceptance have remained unchanged.

A further goal of science is to predict new phenomena, which are then included within the pattern. Both because of the extent of the overall pattern and because extension must come from without the pattern as well as from within, conjectures may be made as to the initial step to achieve extension or better connectivity within the pattern. Conjectures may arise from many circumstances; there are no

199

rules for their proposal. Acceptance, however, depends not only upon the predictions to which they lead, but upon the absence of contradiction or refutation (Popper). Extensions may progress up an ill-defined hierarchy from conjecture through hypothesis, theory, and eventually to Law of Nature. The activity of science is to extend the pattern to include an increasing number of types of phenomena or, in other and more familiar words, to produce and connect Laws of Nature toward that end.

Many would extend the purview of science as enunciated above. We shall not, for this boundary is where we shall differentiate science from technology. Man is the active force behind science; human judgment proposes and accepts the current canons. The goal of the activity is taken as the extension of the pattern, the production of scientific knowledge—no more than that. If science satisfies a human need, it is that related to a desire to extend the scope of the pattern, the desire to understand.

We shall understand technology to be an activity, material or conceptual, directed towards the satisfaction of material desiderata or the material aspect of non-material ones (e.g. artistic or emotional). Knowledge obtained through the activity will be considered technological or technical knowledge. To examine the technological procedure or activity more closely, we shall consider technology as the application of techniques.

Techniques will be taken as the procedures involved in the use of tools to further specific (human) technological desiderata. They may be formulated in an abstract manner; they may consist of a codified or non-codified sequence of physical or mental actions. Their purposeful application constitutes the activity of technology. Demonstration of their variety may be facilitated by example. Thus we may say: Mathematical programming is a technique; the Bessemer Process of steel manufacture is a technique; the methods of striking selected stones together so as to form an axe or to produce fire are techniques; the imposition of a pattern on a series of oscillations so as to convey or transmit information is a technique, as is the production of the oscillations themselves; machine design is a technique; the various types of industrial "know-how" are techniques; food preparation is a technique; etc.

What is called technique here is often referred to as a technology; we shall avoid this usage just as we shall generally refer to science as such rather than the sciences (e.g., chemistry, zoology, etc.) which are analogous subdivisions.

Before going on to define tools, we should pause to note that whereas techniques are applied to satisfy certain desiderata, there may be more than one applicable technique, and one technique may be more satisfactory than another. Technology is goal-oriented; one technique will generally be more efficient or effective than the others. We return later to measurement of efficiency or effectiveness; at this point we introduce them as measures of satisfaction without comment.

Tools will be taken to include conceptual constructs as well as material artifacts. The term will also be applied to those controllable biological parts that act as the environmental interface between an organism and its physical surroundings. In accord with this definition we may consider as tools: any utensil or container; any machine; any instrument or device; the hand, the eye, the voice; mathematics; science; etc.

There may be some question about the inclusion of science and mathematics as tools, or as part of the technological inventory. Moreover, the distinction between tool and technique may also be questioned in some cases. Since our purpose in proposing the definitional set is to test it, however, we shall see how it stands up, and what it may reveal.

To recapitulate: The *tool* is what is applied; the *technique* is how it may be applied; *technology* is the application as well as the knowledge of the technique and the tool. While the products of technological activity depend on the cumulation of prior technology, the desiderata arise within the society itself. The characterization of technology as an extension of man is apt both for the individual and the group. Stone Age man improved his ability to satisfy his basic wants of food and shelter by fabricating such tools as stone axes and simultaneously developed and improved the necessary techniques. The solutions for all persons or all groups were not identical; they were dependent both on the physical and the social environment.

Language, by our definition a technique for communication, may have been developed to meet a psychological desideratum rather than a more material one, but its introduction facilitated exchange of information which favoured increased technological activity.

Group desiderata become more evident with the rise of civilization and the establishment of settled agricultural communities. The desire better to till the soil was met by tools such as ploughs, hoes, scythes, and sickles. The number of desiderata has increased with the complexity of the society within which they arise. We may call the adopted results of the technological activity technological solutions. The tools mentioned above are examples. A technological solution to meet a desideratum is not necessarily unique, however: Scythes and sickles accomplish the same ends. There is a matter of choice to which we shall return.

In the nineteenth century, man was characterized by Thomas Carlyle as a "tool-using animal." More recently, R.J. Forbes has elaborated on man as homo faber (man, the maker), as well as the more usual homo sapiens. We may well speak of "man, the technologist," but we must recognize that technology is not the sole property of man as Carlyle would have led us to believe. Beavers build dams, and birds build nests for their family units. Insects, such as termites, ants, and bees, show technological solutions for their highly organized societal desiderata. It is not our purpose to explore this further: Technology obviously exists on an instinctive level. We shall, however, concentrate on the sapient one associated with man.

Forbes, along with many others, considers control over nature to be the goal of technology as indicated by the title of his work *The Conquest of Nature— Technology and its Consequences* (Praeger, New York and London, 1968). His working definition of technology is similar to ours:

> the mental or physical activity by which man alone, or together with his fellowmen, deliberately tries to change or manipulate his environment.

He goes on to state:

> He (man) may be acting on the ground of empirical or theoretical knowledge or he may simply be following a hunch. Since there is a technological component in virtually every overt action of man there also may be an impulse, subconscious rather than well conceived, to outdo or better this component.

Forbes restricts technology to human action; we shall accept its possible broader scope.

The word *deliberately* used by Forbes corresponds to the role which we have ascribed to desiderata. Whereas initiation of any new technique is normally on the individual level, however, acceptance is on the group level. Without group acceptance the application is inconsequential. Group acceptance of technology cannot be entirely equated to instinctive responses to the environment (e.g., an instinctive desideratum in response to a frigid environment is to provide warmth). But neither is the response or subsequent choice of tool or technique completely dependent on objective criteria. We imply this matter of choice or deliberation, which we take to contain a strong subjective element, not dissimilar to the thematic element which Holton sees in the choice of scientific theory. We should note that if there is no choice but only a conditioned response to the environment, we (society as well as the individual) have no free will. Thus, our examination of technology logically leads to the traditional problem of the existence of free will. As might be expected, this aspect of technology has been

202

recognized in a number of religious critiques, and the traditional recourse to the supernatural is often encountered.

We shall assert that persons and societies definitely respond to their environment on a sapient as well as an instinctive level. We are in no position to define where one response takes over from the other; we shall assume an area of overlap. Perhaps the best justification for such a choice is that acceptance of complete programming (deterministic or stochastic) would destroy the motivation for a study of this type. Even within the context of an analysis limited to technology, we can recognize the critical nature of this assertion or postulate. We may also appreciate the divergent opinions of different authors.

Technology has been taken as a response to the environment. But the response contributes something new. We may agree with McLuhan that "technology creates environments." The process is dynamic; small wonder that we see it change with time. Each increment to technology creates a new world.

Whether the initial desideratum in any case is the result of individual or group action, we may measure our success by the degree to which our chosen technique permits attainment of the specified goal. The new environment created by the application of a new tool or technique may not only meet our initial desideratum, however, but also pose new and unwelcome situations. In fact, it has done so many times. The development of nuclear weaponry is an extreme case in point, but analogous situations, not all military, have occurred throughout history.

We may gain a better insight into the effect of technological innovation on society—i.e., the adoption of new technological solutions—by formulating the situation within the framework of systems theory or operational research. A bounded system is a complex of objects, persons, and activities considered as a first step to change only as a result of internal interactions. When these are known and deemed calculable, the effects of external factors on the system (inputs) and the effects of the system on the environment (outputs) are next evaluated. The method is most effective when inputs and outputs are either readily measured or negligible. It is a most powerful technique.

The problems in the application of systems theory are identical to those of predicting the effects of technology. The system boundary in real life is artificial and is set to facilitate calculation with a minimal distortion of reality. The system model is never a perfect representation because we are never able to consider *all* aspects of the inputs and outputs over an extended time period. Over a short period when inputs and outputs may be negligible or their deviation from calculable values negligible, this system-model approach is adequate. It is the typical approach in engineering where adequate precision and sufficiently short time periods prevail.

While the system approach is less successful in practical application to social systems, it remains a valuable explanatory technique that highlights the intrinsic limitations. We may consider a person or a social group along with its immediate environment as a system, and technological innovation (tool, technique, or application) as either an input, the result of an input, or as a consequence of action within the system (evolution). The formulation places the problem of the effect of technological innovation in a different perspective.

When the new tool or technique is initially applied to meet the new (or improved old) desideratum, we may make our initial calculations within the defined sub-system. These include the degree to which the innovation will meet the desideratum. The short term effect limited to the approach to the specific desideratum may be calculated. This is one of the bases upon which the innovation is accepted. The system model will not contain *all* the factors, however, nor will their effect over an indeterminate time span be precisely known. Hence, although we can predict short-term effects with some confidence, the long-term are intrinsically impossible to predict. The dilemma is clear. Moreover, history shows that the unpredicted side-effects are in the long term more important than the achievement of the particular desideratum.

This phenomenon is a familiar one to systems analysts. The practical solution is to expand the sub-system until a pragmatic boundary is found and maintain an awareness of the time scale of applicability. Although they have not formulated the problem in the above terms, a number of critics of technology have proposed that systems analytical practices be used to control technology without recognition of the intrinsic limitations. By control of technology, they mean that not only should a moral as well as a material value be placed on the desideratum or goal, but that all subsequent effects on the external environment be calculated and weighed against the worthwhileness of the desideratum. This recommended course of action implies an ability we have not yet been shown to possess except in short-term limited situations.

As the time dimension is automatically introduced in the recommendations for control, we can legitimately turn to history for illustration. Despite the number of modern illustrations, we take five historical ones from the Renaissance and earlier because both their side-effects have persisted to the present, and they are clearer in historical perspective.

Lynn White, Jr. has built two impressive illustrations. In the first a tool, the metal foot-stirrup, made possible a change in mounted combat by giving the rider lateral stability. This made profitable the use of new tools, lances and armour, and a whole new combat technique. The combat technique, however, required a heavy horse, the destrier, to carry the armoured cavalier. To achieve this, Charles Martel in 732 A.D. adopted a system of pasture and land ownership that evolved into the Frankish and subsequent West European feudal system.

White's second example starts again with a tool, the heavy plough with mould-board, which made it practical to till the heavy, rich soils of Northern Europe. The plough required more powerful prime-movers (multiple ox and horse teams) than were economically or otherwise available to the individual farmer. Moreover, the most efficient technique was a method of ploughing in long oval areas. Both the necessary cooperation and the shape of the fields fostered the manorial system.

Carlo Cipolla gives a Renaisssance example linked to the introduction of cannon (the technique for whose production was developed by medieval bell casters who wished to ring out the greater glory of God) along with their use on the newly developed square-rigged sailing vessel. The combination permitted an armed sea-borne traffic which European powers exploited to obtain colonial outposts in the Far East. The repercussions abound to this day.

Marshall McLuhan and others select the introduction of the phonetic alphabet in antiquity as a key tool or technique which, when combined in the fifteenth century with the techniques of inkmaking, paper manufacture, and metal casting, not only led to the communications explosion of books, but to the ascendency of linear thinking which combined so well with the printed word. From this thesis McLuhan expounds new effects and changes arising from the new non-linear medium of television. Whereas the first three examples relate to effects on society, this latter indicates a thoroughly unintentional and unexpected effect on the mode of human thought.

Lewis Mumford, who uses the term "technics" with a meaning including our terms technology and technological knowledge, has a remarkable example which he traces back to the Bronze Age, if not the Neolithic: the introduction of the *technique of organization*, the arrangement of components into a machine which in turn is a tool to accomplish a desideratum for which the individual components are inadequate. There have been many who have joined a philosophy of machines to a study of technology, but Mumford's great machine, the *mega-machine*, is unique because the components are human. It is the use of the tool-technique of the mega-machine which Mumford claims made possible such feats as the pyramids of Egypt, the Great Wall of

China, and many military applications. The subsequent history of technology may be considered largely as the replacement of the human components with material ones and of human energy with alternate, more efficient sources.

Before returning to Mumford and efficiency, we should examine the interpretation of Jacques Ellul. In *The Technological Society* (Vintage Books, New York 1964), Ellul considers technology to have affected the nature of thought, but differently than McLuhan does. In our terminology we may say that one common desideratum has been the better, simpler, easier accomplishment of a second desideratum. We have implied that these descriptive adjectives may be subjective; they may, however, take an objective formulation as is done, say, in the measurement of the efficiency of heat engine. In Ellul's opinion Western society, at least, has unconsciously substituted the objective measure for the subjective one. The role of machine efficiency in Ellul is similar to that which McLuhan assigns to print in which *linear thinking* is virtually forced upon one when communication is in print or phonetic symbolism. Other media lead to other results, thus his well-known and cryptic "The medium is the message." The adopted desideratum of an objectively expressed efficiency, a mathematical tool, is effectively a grant of autonomy to the tool or technique, so that the automatic human (or is it inhuman?) response is such that other possible desiderata are not considered and the technological machine rides over man's potential humanity. We need not accept Ellul's pessimism. We must, nevertheless, agree that he has expressed a frightening prospect; technology as a virtual monster whose Frankenstein we are. In this perspective few things exceed technology in importance.

Mumford relates technics to humanity in a somewhat different but equally critical manner. In his earlier works he did not apparently consider the danger—and to Mumford it is a real danger, for he is a strong proponent of the dignity of man and hence his autonomy—that man might through his "free" adoption of technology become effectively programmed. He introduced the well-known descriptions Eotechnic, Paleotechnic, and Neotechnic to apply to historic stages in the development of technology. Initially, Mumford was optimistic that these technics would serve to liberate man; to him the machine was at first man's tool. Latterly, he has recognized more clearly the technique (our term) of organization that pervaded technology (our definition) from antiquity. It was at this point that he discovered the megamachine.

Mumford, unlike Forbes, recognizes animal as well as human technology—most probably because he refuses to place the essential

206

characterization of man as the unique possession of a non-spiritual attribute. He believes man's unique essential qualities are of a different and certainly non-material nature. We are not solely homo faber or even homo ludens. He claims the introduction of organization—the machine—came from the associated rhythms and division of labour accepted better to facilitate ritual and a communion with gods and nature. Thus, the mega-machine initially was adopted freely for humanistic rather than material purposes. That it also facilitated material goals was incidental, but in its adoption for these purposes we may have misdirected human progress, with organization as a goal not a means. Underlying Mumford's concern, and that of many others, is the fear that technology will pervert the human quality of man. Thus technology, both unnatural and inhuman, is a danger to him as to Ellul.

We may share their concern, for it is no desideratum of this author's that man be denigrated. Unless we wish to take recourse in the supernatural, introduce a new Manicheanism, or to separate Man from Nature, however, it is difficult to consider technology as unnatural. And we may well recognize that technology is vital for the survival of the vast majority of mankind and has been a prime factor in the historical evolutionary pattern of society. Whether or not technology is part of the essence of man, he has *acted as if* he were homo faber, perhaps even more successfully than he has acted as homo sapiens. We continue here to consider technology an extension of man himself. We shall not explore the paradoxical question of man's inhumanity to himself: This is beyond our province.

As long as we are in control, we may use technology for attaining *our* desiderata. If we recognize that we may lose or have lost control, the knowledge may help us to retain or regain it. Ellul sees this clearly in his proposed solutions. If, however, we have become a new "homme machine" with conditioned habits and conditioned or programmed minds, we are in a predicament from which we can gain only the slight cheer that we perceive it. Technology (Ellul's Technique) is a sharp tool; we take it, however, as a human one.

As a human tool, technology has been explicitly connected with the idea of progress—at least since Francis Bacon, whose Utopian *New Atlantis* is one of the earliest of a series of technologically influenced Utopias which continue to the present day. Perhaps the best known of the modern Utopians is R. Buckminster Fuller. One of his several definitions of technology (a restricted one; he calls it realized technology) is the "objective uses of all that man had found out about his physical universe in chemistry, physics, mathematics, geography, geology, biology, and other studies." He has his doubts about the choice we shall make, but he has no doubts about our potenitial to make that choice.

Technology or machines, however, are not necessarily part of all Utopias since Bacon. Far from it: The bucolic dreams of a "Golden Past," Rousseau's "noble savage" (non-technical, of course), and an agrarian ideal perturbed by a "machine in the garden" co-exist and have co-existed. The fear of the machine is brought out strikingly in Samuel Butler's *Erewhon*.

Science fiction as a serious genre has explored technological potential in a series of contrived make-believe societies. While horrors are often brought out, the underlying feeling, while not exactly optimistic, seems to accept technology as human and as a road to a better life. Ben Bova, currently editor of a major science fiction journal, *Analog*, sums it up by saying that man without technology is not a noble savage, but a starving, naked ape.

From the preceding it should be clear that there is in our terms no place for a discussion of a modern topic or pseudo-topic: the morality of technology. Technology has no volition and no ability to make moral judgments. This is not to say that it cannot pose a moral problem, even one apart from the moral judgment we assert in establishing our desiderata and making our choices. This lies in the long-term unpredictability of the outcome of application. For those who believe this moral dilemma to be a recent discovery—it obviously is not of recent origin—it may be interesting to read the discussion in Agricola's *De Re Metallica* (1556).

We have seen the limitations of trying to predict the long-range effects of technological innovation; we must also note the futility of trying to isolate technology as a bounded system for study. The technological process affects every aspect of society, even man's thought processes. But we can examine the inputs and outputs which relate technology to other aspects of society.

There is no doubt that the body of scientific knowledge is today the most important tool of technology. It is not the sole tool, however. Artifacts and machines play an important and obvious role. It is difficult to find a serious student of philosophy of technology who will make the bald statement, "Technology, as it is today, is nothing but applied science." In common usage, however, we find the implication frequently. This cannot be dismissed as insignificant. The words "science" and "technology" are so linked in daily use that a recent Canadian government report on science policy, despite its title, devotes as much time, if not more, to what we (and they) would call uses of technology. Even the more cautious and scholarly work of Mario Bunge has been summarized by Wm. Kuhns as viewing technology as applied science.

Such an attribution is unfair. Bunge develops a different definitional framework than we do here: He differentiates types of technological theoretical concepts in systems formulations. Bunge focusses on one specifically human role, that of the decision maker, who delineates the desideratum and makes the decision for action. Unless brainwashed, he is clearly in command. Bunge differentiates a substantive scientific theory (his example is hydrodynamics) from a substantive technological one (hydraulics). He considers "substantive technological theory to be based on scientific theory and to provide the decision maker with necessary tools for planning and doing." He defines operative technological theory (his example is queueing theory) as that directly concerned with action or adoption. His example of an operative technological theory and ours of a technique (mathematical programming) are identical in essence.

Bunge's analysis is not antithetical to ours. It could well serve to influence our next iteration. His concern seems to be that some philosophers have been misled into thinking that scientific theories are *nothing but* tools. This apparent direct blow at our terminology is not as direct as it reads. We can reiterate that scientific theories or science are *tools* to the decision maker without claiming that they are tools to another scientist. We concur with Bunge that scientific theory may undergo "conceptual impoverishment" as a technological tool. In fact this fits ideally into our own conceptual structure of technology. Most of our disagreements with Bunge lie in definition.

The importance of definition should by now be apparent. It reflects the philosophy of the proposer, his special expertise, even his or her prejudices. Nowhere is this more clear than in the Marxist influenced analyses. Whereas some Marxists have recognized the theoretical possibility that the goal of science could be the extension of a pattern of knowledge, as we have taken it to be, they believe the purpose of the extension should be for human betterment through application. Since much technological innovation in the past was associated with a capital outlay, and thus controlled by those who possessed the capital, it is claimed—with some justice—that the innovations were selected with the additional desideratrum on the part of the decision maker to improve his personal position and that of his class. The Marxist solution is to place this power of choice in the hands of the people, primarily the proletariat. They believe that technology, so directed, may lead to the better social state. Since they take science and technology to have the same goals, differentiation is not made. Both are part of the historical evolution of society.

There are other reasons for confusion: Much scientific extension has only been made with the aid of devices deemed technological—the telescope, the microscope, etc. The construction of certain complex scientific apparatus—e.g., atom smashers—is an answer to the non-material desideratum of extending science although with a proximate technological one of building the device. It thus depends on one's orientation whether one says that a space laboratory is a scientific achievement or a technological one. In itself the difference in word choice is of no consequence unless it causes confusion. And unfortunately, it has.

Within our preliminary definitional framework we have examined modern and older views of technology. We see it as a part of a historical process; we see its interaction with society, its dependence on man (unlike science); we see the limitations of our control. For these purposes the framework is adequate.

Notes on Contributors

JOHN W. ABRAMS received his Ph.D. in astrophysics at the University of California, Berkeley, in 1939 after having done his undergraduate work at Berkeley and further graduate study at Harvard, Leiden, and Tartu (Estonia). He instructed at the University of California until joining the Royal Canadian Air Force in 1940. He served as a navigation instructor and later as an operational research officer (Squadron-Leader) with the RAF in England. He taught physics at the University of Manitoba in 1945-1946, and astronomy at Wesleyan University (Connecticut) 1946-1949. He returned to Canada in 1949 to the Defence Research Board serving in Ottawa, London (England), and Paris, where he was Chief Operational Analyst to the Supreme Headquarters Allied Powers Europe 1958-1961. He left the Defence Research Board where he was Chief of Operational Research in 1962 to join the University of Toronto faculty. While in London as Adviser to the Admiralty (1949-1951), he also studied History and Philosophy of Science at University College. He was Scientific Adviser to the Chief of Air Staff, RCAF (1955-1958). At Toronto he was founding director of the Institute for the History and Philosophy of Science and Technology (1967-1972) and is currently professor of history and professor of industrial engineering. He is on the Council of the Centre for Technology and Culture at the University of Toronto, the Executive Council of the Society for the History of Technology and is Secretary-General of ICOHTEC (The Internation Cooperation in the History of Technology Committee) and past-president (1971-1974) of the Canadian Society for the History and Philosophy of Science. In 1975 he was a Visiting Professor of Philosophy at the University of Victoria. He has also taught at Queen's University (Kingston, Ontario), Trinity College (Hartford), and the University of Aston (Birmingham) and lectured at American, European and Canadian universities. His current interest and research field is in the history of philosophy of science and technology. His principal publication field is operational research. His latest paper in the philosophical area is "The Canons of Scientific Acceptability" in the volume *Tommasso d'Aquino nel suo VII Centenario*, Rome, 1974.

MARIO BUNGE was born in Buenos Aires in 1919. He founded and directed a workers' school (1938-1943) and edited *Minerva*, a philosophical magazine, (1944-1945) while studying theoretical physics. He obtained his doctorate in physics from Universidad Nacional de La Plata (1952) and published a number of articles on atomic and nuclear physics, as well as on the foundations of quantum mechanics and the philosophy of science. He was an instructor in physics at the Universidad Nacional de Buenos Aires (1947-1952) until he was dismissed for refusing to join the Peronist party. After the fall of that dictatorship, Dr. Bunge was appointed an assistant professor, and shortly thereafter a full professor of theoretical physics at the Universities of Buenos Aires and La Plata (1956-1959). In 1957 he won the chair of philosophy of science at the University of Buenos Aires, which he resigned in 1963. From that time on, he held visiting professorships in philosophy and physics (Temple University and University of Delaware), and an Alexander von Humboldt research fellowship in Germany (1965-1966). In 1966 he joined the Department of Philosophy at McGill University. He has also been a visiting professor at the universities of Pennsylvania, Texas, Freiburg, México, and Aarhus, as well as at the E.T.H. in Zurich. Professor Bunge's philosophical interests include the philosophy and methodology of science, in particular of physics and sociology, what he calls the semantics and the metaphysics of science, and value theory and ethics. He also continues to be interested in the axiomatic foundations of physics (in particular quantum mechanics) to which he has recently added a dabbling in the foundations of mathematical sociology. He is currently engaged in writing a seven volume work that will expound his philosophical system, which he hopes will be fairly exact as well as consonant with science. His publications comprise more than 200 items in twelve languages. His main books are *Causality* (Harvard University Press, 1959), *Metascientific Queries* (Charles C. Thomas, 1959), *Cinemática del electrón relativista* (Universidad Nacional de Tucumán, 1961), *Intuition and Science* (Prentice-Hall, 1962), *The Myth of Simplicity* (Prentice-Hall, 1963), *Foundations of Physics* (Springer-Verlag, 1967), *Scientific Research*, 2 volumes (Springer-Verlag, 1967), *Philosophy of Physics* (Reidel, 1973), *Method, Model and Matter* (Reidel, 1973), *Sense and Reference* (Reidel, 1974), and *Interpretation and Truth* (Reidel, 1974).

ROBERT E. BUTTS has been Professor of Philosophy in the University of Western Ontario since 1963 where he was chairman of the Department of Philosophy for nine years. Professor Butts studied at Syracuse University (B.A. magna cum laude, 1951; M.A. in philosophy, 1952), and at the University of Pennsylvania (Ph.D. in philosophy, 1957). In 1962-1963 he was a National Science Foundation Postdoctoral Fellow in History and Philosophy of Science at Trinity College, University of Cambridge. A specialist in problems of scientific methodology, the philosophy of Kant, nineteenth-century British philosophy of science, and the history of philosophies of science, Professor Butts has taught at the University of Pennsylvania, St. Lawrence University, and Bucknell University. He has held visiting appointments at the University of Calgary, University of Illinois (Urbana), and University of Pittsburgh. Professor Butts has published over 35 items on Kant, the history of philosophy, William Whewell, and contemporary problems in philosophy of science. He has edited or co-edited *William Whewell's Theory of Scientific Method* (Pittsburgh 1968), *The Methodological Heritage of Newton* (Toronto & Oxford 1970), and *Science, Decision and Value* (Dordrecht, Holland 1973). In 1974 he was commissioned to write "Philosophy of Science in Canada," by *Zeitschrift für Allgemeine Wissenschaftstheorie*, v. 2. A Polish translation of the work will appear in Vol. 33 of *Ruch Filozoficzny*. Dr. Butts is now working on a book on Kant's philosophy of science, and is co-editing *New Perspectives on Galileo* and the *Proceedings* of the fifth International Congress of Logic, Methodology and Philosophy of Science. He was chairman of the Local Organizing Committee of this congress, the first to be held in Canada. In addition to his publications, Dr. Butts has presented over 40 papers and lectures, including nine lectures delivered during a lecture tour of Scandanavia in 1973. Professor Butts serves as consultant to a number of publishers and granting agencies and is on the editorial boards of a number of journals and series, including the University of Western Ontario Series in the Philosophy of Science, which he helped to establish. He has served on committees of the Canadian Philosophical Association, the American Philosophical Association, and several international bodies. He is now second vice-president of the Division of Logic, Methodology, and Philosophy of Science in the International Union of History and Philosophy of Science.

FRANK CUNNINGHAM studied philosophy and social theory at Indiana University, the University of Chicago, and the University of Toronto. He is currently an associate professor of philosophy at the University of Toronto, where he has been teaching courses in the philosophy of the social sciences, Marxism, and related subjects since 1967. He is co-secretary of the Society for the Philosophic Study of Marxism, Western Division (an organization of the American Philosophical Association), an executive of the Committee on Socialist Studies (a member of the Canadian Learned Societies), and is on the editorial board of the journal, *Philosophy of Social Science*. Published works by Professor Cunningham which relate to the subject of the essay in this volume are *Objectivity in Social Science* (Toronto, 1973) and "Practice and Some Muddles about the Methodology of Historical Materialism," *Canadian Journal of Philosophy*, vol. III, no. 2 (December 1973), pp. 235–248. The present essay is based on an earlier one, "Marxism and Social Science," in *Communist Viewpoint*, May–June, 1975, pp. 37–44.

THOMAS A. GOUDGE was born in Halifax, N.S. and was educated at Dalhousie University, University of Toronto, and Harvard University. He has taught at Waterloo Lutheran University, Queen's University, and the University of Toronto, where he served as Chairman of the Department of Philosophy from 1962–1969. During World War II he was in the R.C.N.V.R. Professor Goudge was elected a Fellow of the Royal Society of Canada, President of the Charles S. Peirce Society, and later President of the Canadian Philosophical Association. At different times he has been a member of the Canada Council's Academic Advisory Panel, of the Council for Philosophical Studies, and of the editorial boards of *The Encyclopedia of Philosophy* and *The Monist*. On two occasions he has held leave fellowships from the Canada Council. His publications include: Bergson's *Introduction to Metaphysics* (edited with an Introduction, 1949; 1955); *The Thought of C. S. Peirce* (1950; 1969); *The Ascent of Life* (1961—Governor General's Non-fiction Award; rep. 1967); *De Wijsgerige van de Evolutie* (1966; Dutch tr. of *The Ascent of Life*). Other publications include more than 50 articles in philosophical journals and encyclopedias. These are mainly devoted to epistemology, philosophy of mind and philosophy of biology, but some have to do with the history of nineteenth-century and early twentieth-century philosophy. Recently he has prepared two chapters on Canadian philosophy 1910-1973 for the revised *Literary History of Canada* which is in press.

JOHN O'MANIQUE was born in Ottawa in 1936, is married and has five daughters. He received a B.Sc. in Mathematics and Physics from St. Patrick's College in 1958, studied Philosophy at the University of Ottawa (Ph.D., 1966), and did post-doctoral research in philosophy of science at Cambridge University in England. Dr. O'Manique has published several articles on Teilhard de Chardin, as well as *Energy in Evolution* (Garnstone Press, London and Humanities Press, New York, 1969). Since 1968 he has been involved in interdisciplinary programs at St. Patrick's College, Carleton University where he is an associate professor of philosophy; he is a member of The International Center for Integrative Studies. Recently, Dr. O'Manique has been working on the ethical and behavioural dimensions of ecological problems as an advisor to the Third Generation Research Project for the Club of Rome, and as a consultant to the Canadian Department of the Environment. His publications in the area include "The Ethics of Survival" in *Philosophy Forum* and "Values for Survival" in *Values and the Quality of Life*. He is also continuing to study models of consciousness in the context of contemporary science.

ALEX C. MICHALOS is a Professor of Philosophy who has taught at the Univeristy of Guelph, Ontario since 1966. He has written *Principles of Logic* (1969), *Improving Your Reasoning* (1970), *the Popper-Carnap Controversy* (1971), and recently *Foundations of Decision-Making*, as well as several journal articles and book reviews. He has edited *Philosophical Problems of Science and Technology* (1974) and co-edited a volume of the series *Boston Studies in the Philosophy of Science*. He is the Founder and Editor of the journal *Social Indicators Research*, and a member of the Editorial Boards of *Dialogue* and *Theory and Decision*. For the past five years most of his research time has gone into a two volume study of the quality of life in Canada and the USA from 1964 to 1974—which is probably where most of the next four years will go too. The present essay, "Decision Making in Committees," is based on an article that appeared in the *American Philosophical Quarterly*.

ALEXANDER ROSENBERG is a member of the Department of Philosophy at Dalhousie University in Nova Scotia. He teaches in the philosophy of physical and biological science and the philosophy of social and behavioural science, as well as a course on Hume's *Treatise* and Kant's *Critique of Pure Reason*. Rosenberg pursued his graduate work at Johns Hopkins University, and has been a visiting professor at the University of Minnesota and its Center for Philosophy of Science. He has published papers on the conceptual issues in the explanation of mental phenomena by computer simulation in *Philosophy of Science*, on questions in the philosophy of social science, and economics in particular, in *Philosophy of Social Science, Theory and Decision*, and *The American Political Science Review*, and on causality in *Journal of Philosophy, American Philosophical Quarterly, Philosophical Studies, Philosophical Forum*, and *The Personalist*. He has contributed papers on these subjects to anthologies, including *Philosophical Problems of Causation* (ed. T.L. Beauchamp) and *Developments in the Methodology of Social Science* (ed. W. Leinfellner and E. Köhler) and on more general questions in the philosophy of science in *Dialogue* and *The Canadian Journal of Philosophy*. Rosenberg is also the author of a book on methodological and conceptual issues in economic theory, *Microeconomic Laws: A Philosophical Analysis*, published by the University of Pittsburgh Press.

MICHAEL RUSE is 35 years old, was born in England, and educated at the University of Bristol, England; McMaster University, Canada; and the University of Rochester, U.S.A. He has taught at the University of Guelph for ten years and his interests are in the history and philosophy of science, particularly biology. He is the author of some 36 papers on various philosophical and historical aspects of biology, and of a book published in 1973, the *Philosophy of Biology*. At present he is writing one book on Darwin and the revolutionary effect of his thought on nineteenth-century science, religion, and philosophy, and another book on ethics and the life sciences.

TOM SETTLE's earliest research concerned the impact of science on theology, and especially the challenge posed for theology by the standards of rationality and of rational integrity adopted in science, and then, conversely, the challenge posed to popular views of science by the discovery that the rationality of its methods does not guarantee the truth of its findings. He has expounded, developed, and criticized the philosophy of Sir Karl Popper whose work he thinks deserves the greatest respect but who is often misunderstood or misrepresented and thus criticized only in caricature. Popper's critical (or skeptical) approach to philosophy has been extended by Settle into the domains of political economy and rational ethics. Two directions of research come together in the book *In Search of a Third Way*: first, science policy where the moral dimension of scientific and technological research seems inescapable and second, the philosophical presuppositions of science. Assuming a non-justificationary approach to knowledge, speculation in metaphysics—including theory of man—becomes permissible if it is held open to criticism. Tom Settle has taught philosophy at the University of Guelph in Ontario since 1967 and is currently Professor of Philosophy and Dean of the College of Arts there.

WILLIAM R. SHEA was born in Gracefield, Quebec, in 1937. He was educated at the University of Ottawa, the Gregorian University in Rome, and Cambridge University where he was chairman of the Students' Association of Darwin College and obtained his Ph.D. in philosophy. Before going to McGill, where he is the chairman of the History and Philosophy of Science Program, Professor Shea taught at the University of Ottawa, and was a Fellow of the Renaissance Center of Harvard University, a Senior Fellow of the Canadian Cultural Institute in Rome and a Visiting Professor at the Institute for the History of Science in Florence. Professor Shea is the chairman of the Canadian National Committee for the History and Philosophy of Science and

214

a consulting editor of *Dialogue*, the journal of the Canadian Philosophical Association. He contributes regularly to philosophical and historical journals, and he is the author of *Galileo's Intellectual Revolution* (New York and London, 1972), and the co-editor of *Reason, Experiment and Mysticism in the Scientific Revolution* (New York and London, 1975). Professor Shea's main interest is the history and philosophy of science from the seventeenth century to the present, and he is preparing an English translation and commentary of Descartes' *Le monde*. He is also collaborating, with his colleague Jonathan Robinson, on a translation and commentary of Hegel's *De Orbitis Planetarum* on the nature of science and the orbits of planets.

JAMES VAN EVRA was born in Chicago and received his early education in its environs. He holds a B.A. from Valparaiso University (Indiana) and an M.A. and Ph.D. from Michigan State University. After teaching briefly at Michigan State, Professor Van Evra accepted a position at the University of Waterloo in 1965. He is currently an associate professor and acting chairman of the Philosophy Department there. Professor Van Evra is book review editor for *Philosophy of Science,* the journal of the Philosophy of Science Association. He served on the programme committee for the 1974 biennial meeting of the Association. Professor Van Evra is co-editor of *PSA 1974* (Reidel, forthcoming). He is the author of articles on the history of logic and the philosophy of the behavioural sciences.

Date Due
